ADVANCE PRAISE

"Having been in retail, food service, and hospitality for over 47 years – where the actions of the front lines associates can define the success of the business, where communication is vital for product quality, and where through-put is directly related to the decision making of those closest to the end user – I read with enthusiasm Kevin Snook's book and wished I had another run in me... in the manufacturing sector!

"In his own way, Kevin Snook touches on all of the points above, and more, in his book, *Make it Right*. Kevin's focus is on manufacturing, but his five points around the ALIGN process apply to any business sector, organization, individual business department, and even your own family.

"I have made the statement for years that being successful in business is not about 'rocket science' – it has to do more with 'social science.' Kevin's book shows that to be true."

–JIM DONALD, *Former CEO of Starbucks, Pathmark Supermarkets, Haggen Food & Pharmacy, and Extended Stay Hotels*

"As a client, and one of the case studies for *Make It Right*, I can attest that Kevin practices the principles he explains in this book. This step-by-step guide to aligning the business helped the manufacturing division of DSG International produce efficiently across three countries during a challenging and sustained period of 35% annual growth. Real advice, real results."

–AMBROSE CHAN, *CEO, DSG International Limited*

T0163747

"Kevin is a resounding professional who has distinctive competence in the arena of manufacturing and operations. I have personally worked with him in setting up a state-of-the-art low-cost site for shampoo in a record period of five months, and seen how he leads from the front. Kevin believes in getting the best from the people and making his dream for a project a shared dream. He is non-stop open to refinement of the path ahead, even if ideas about how to do that come from the junior most person on the site.

"Given this bandwidth of passionate experience, I am glad that Kevin has decided to go ahead and share his pearls of wisdom and experience in *Make it Right*. I see it is as a compelling read for anyone who has anything to do with manufacturing and operations, at the least."

–VAGISH DIXIT, *Managing Director, ALPLA India Private Ltd. (ALPLA is a US$3.5 billion global major in the field of manufacturing of plastic bottles and caps, with a global footprint of more than 160 plants in 42 countries)*

"Most books are written about what manufacturing is from the outside. This book is about what manufacturing actually *is*, and what companies should really *do*."

–SANDESH AGGARWAL, *CEO, Leo Plastics Ltd, Tanzania*

"Giving 'inspiration' to our employees is where we have made most progress, not only with manufacturing managers but also with operators in the factory. Mr. Snook directly communicates with them and helps them think of ideas that could make immediate improvements."

–H. S. CHOI, *COO, KleanNara Company Ltd.*

"If you have ever felt a lack of alignment between what you expect and what your teams deliver, then *Make It Right* is a book for you. It is filled with practical and easy to implement actions that you can immediately begin to use to close that gap."

–RISHI KHEMKA, *Chief Enjoyment Officer MindBox, www.mindbox.in*

"In my profession, I've met many professionals across multiple manufacturing industries around the globe for the past 30 years. However, seldom have I had the pleasure to work alongside someone as inspiring as Kevin Snook. His analyzing skills and working style ensure that all stakeholders involved are participating actively in the process to *Make It Right*!

–STEFAN VAN DER SLUYS, *CEO, B.G.L Co. Ltd, www. bestgloballogistics.com; President of the Netherlands-Thai Chamber of Commerce*

"The data and methods in *Make It Right* cannot be argued with and give us clear direction for priority setting and planning."

–B. M. CHOI, *Chairman and CEO, KleanNara Company Ltd.*

"*Make It Right* is a must-read for anyone who is committed to the deepest level of success, that place where your life's work becomes an instrument for transformation in your business and in your life. Kevin Snook is a master catalyst for inspiring, teaching, and leading a revolution from within that transforms us and everyone we touch. The world needs us all to *Make It Right*!"

–DR. JAMES ROUSE, *Founder, Healthy SKOOP, Optimum Wellness Media, and Well and Company; he is an entrepreneur, physician, athlete, father, and husband*

"This quote from his book best captures the spirit and energy of Kevin Snook: 'Having helped transform the lives of thousands of frontline decision makers, I still don't feel that I have done enough.' Kevin has found a way to inspire in those he meets a level of conviction, meaningful engagement, and sustainable motivation that goes beyond the metrics of the business.

"Re-tooling one's mindset can lead to transformational leadership that is applicable in even the coldest, most production-driven cultures and environments. *Make It Right* shares five simple steps to help you achieve results with your operations team. Kevin's leadership and experience in applying these principles with the front line have consistently proven the direct relationship between the frontline and bottom-line results – guaranteed!"

–DAVID PELLETTIER, *Top Producing Realtor with RE/MAX*

"Kevin Snook's book on managing the manufacturing business towards 2020 reads like a really good cookbook written by a master chef: You just can't put it down, and you know that with that level of expertise to advise you, you're going to be able to create something extraordinary.

"What makes *Make It Right* extraordinary? The human to human approach. The book was written by someone who cares and wants to help. His style of narrative makes a huge difference – you are not simply reading a book; you are with your personal coach, who knows your business and is committed to helping you. This book is highly practical and is focused on today's manufacturing industry. It's a book to be used as a guide for the leaders of tomorrow. *Make It Right* is not only about managing efficiently for today. It is also about leading your manufacturing enterprise into the future."

"*Make It Right* is refreshing, inspiring, practical, and results focused. Human to Human. Thoroughly recommended."

–JOANNA KALKSTEIN, *CEO, Kalkstein Pty Ltd,*
www.joanna.kalkstein.com

"Kevin Snook takes servant leadership to a whole new level. His ALIGN concept defines the practical process for working with frontline employees to optimize results and create an improved work place. *Make It Right* is full of useful gems of wisdom that every leader can apply. His book provides a fresh and insightful approach that will be helpful to all manufacturing and business leaders."

–RICK STUIT, *Engineering Director,*
Johnson & Johnson APAC

"*Make It Right* is a must read from the CEO down to through every individual in the organization who wants their operations to be world class. It was written for everybody to understand, even those with no experience in manufacturing processes. Thank you for putting your beliefs, skills and passions into this book. It was very inspiring indeed and I read it from cover to cover in one day! I have been in production and maintenance for more than three decades and this book has given me new perspectives and ideas I can use in my current work as an engineering and maintenance consultant."

- RENE NORIEGA, *Maintenance and Project Management Consultant, Vibelle Mfg. Corporation*

"Kevin Snook's extensive experience and wisdom, shared in this book, will benefit everyone tremendously. As an author who reads a lot of books, *Make It Right* is one of those great books that you will want to finish in one go and re-read again and again."

- PAUL PATTARAPON SINLAPAJAN, *Author, Investor, Financial educator, Entrepreneur*

MAKE IT RIGHT

MAKE
IT
RIGHT

5 STEPS TO ALIGN YOUR MANUFACTURING BUSINESS FROM THE FRONTLINE TO THE BOTTOM LINE

KEVIN SNOOK

NEW YORK

LONDON • NASHVILLE • MELBOURNE • VANCOUVER

MAKE IT RIGHT

5 Steps to Align Your Manufacturing Business from the Frontline to the Bottom Line

Published in New York, New York, by Morgan James Publishing in partnership with Difference Press. Morgan James is a trademark of Morgan James, LLC. www.MorganJamesPublishing.com

The Morgan James Speakers Group can bring authors to your live event. For more information or to book an event visit The Morgan James Speakers Group at www.TheMorganJamesSpeakersGroup.com.

ISBN 9781683506706 paperback
ISBN 9781683506713 eBook
Library of Congress Control Number: 2017911261

Cover and Interior Design by:
Chris Treccani
www.3dogcreative.net

In an effort to support local communities, raise awareness and funds, Morgan James Publishing donates a percentage of all book sales for the life of each book to Habitat for Humanity Peninsula and Greater Williamsburg.

Get involved today! Visit
www.MorganJamesBuilds.com

DEDICATION

For the thousands of frontline manufacturing employees I meet on my travels around the world. Without these dedicated and far too often unrecognized individuals, there would be no production.

TABLE OF CONTENTS

INTRODUCTION

———◦———

"I don't spend my time pontificating about high-concept things; I spend my time solving engineering and manufacturing problems."
Elon Musk

Manufacturing is challenging – and only a small percentage of companies do it well. For executives who have come through the ranks of supply chain and manufacturing companies, the challenges can seem endless. For those leaders brought up in sales and marketing functions, the additional complexities of manufacturing can be extremely frustrating, often raising the question, "Why don't we just outsource this?" For those who *have* outsourced, there is quickly a realization that this is not the end of the issues – the issues have simply been transferred.

Manufacturing is complex. While the specter of missing shipments and unavailability of product is costly, the potential for a major market recall due to a quality issue, or a safety incident at the factory, can keep us awake at night. Quality and safety issues can be career-ending for the leader but possibly life-ending for the consumer or employee.

The responsibility on the shoulders of the manufacturing company CEO or senior leader is huge, and there is no way around this. Managing the complex interactions between people, machines, and materials is often treated like a black art, with leaders learning through trial and error.

There are very few schools offering training in manufacturing leadership and, due to the competitive environment, there is even less sharing of best practices throughout the industry. Hence, companies and their leaders learn by doing, and many factories end up looking like high school projects – cobbled together over the years by well-meaning but untrained employees. As the business grows, so does its complexity. With the constant demand for more volume, without the time allowed for realignment, the high school project continues, building on a shaky foundation.

Many manufacturing CEOs are also business owners. Whether taken on through starting or inheriting a business, or as a shareholder, ownership drives a desire for long-term success and the building of a legacy. The future reputation of the business is in the hands of the leader, and the required drive for continual growth never allows the pressure to be released.

Business growth primarily comes through sales, searching out new markets, innovating, and maximizing capacity to keep the costs to the consumer as low as possible, while driving value. Each of these factors puts extra pressure on the manufacturing division, and this creates additional risk.

The CEO, by continually driving the business forward, inevitably creates his own manufacturing headaches. When I walk the factory floor with business leaders around the world, there are conditions and behaviors on show that they are far from proud of; there is a gap between their expectations for the business and the actual situation.

Of the hundreds of business leaders I have supported, they all mention a common challenge: the people. High-specification machines can be acquired, state of the art factories can be built, and high-quality raw materials can be purchased, but questions like these consistently arise about the people involved in the business:

- "Why don't the factory workers just do things the way we want them to?"
- "Why can't we find good people, and, when we do, why can't we keep them?"
- "Why doesn't the factory floor follow the Standard Operating Procedures?"
- "Why can't we change the minds of the workers?"
- "Why does everything need to be challenged?"

Without the factory employees making the right decisions every minute of the day, the best new equipment, an ideal work environment, and quality materials are wasted. The manufacturing employees hold the reputation, profitability, and overall success of the business in their hands, and they are likely to be the most diverse group of individuals in the company.

With the CEO leading the entire organization, dealing with sales, marketing and finance in addition to all the other corporate responsibilities, the role of manufacturing leadership is often delegated. Because the chosen leader often comes from a different industry, or from a different corporate culture, delivering a consistent and appropriate message to the workers is a challenge. What is the "right" manufacturing system to follow, and how will the leadership style of each individual impact it? Even leaders with great backgrounds can get stumped when trying to operate across businesses and cultures. Continuing to change leaders until one finally "gets it right" is a costly process for everyone in the organization and, as you may know, trying to pick the right leader to hire is akin to buying a lottery ticket.

Then we have the flocks of management consultants ready to prey on any sign of weakness. These groups of highly paid consultants, who prefer to work in corporate offices with a Starbucks coffee shop downstairs, are often completely out of their depth in an out-of-town manufacturing facility. The consequence of this is that a team of 27-year-old lean, six-sigma "experts" is invited to the factory to come up with a strategy for a manufacturing process they don't actually know enough about. Armed with hundreds of manuals and the mandate to customize a standard process for your situation, it is not surprising that the vast number of these turnaround efforts fail once the initiative hits the factory floor. Although the concepts of *lean* and *six-sigma* are sound in theory, the challenge of practical application, including the discipline required to carry it off, results in a low success rate when using management consultants in this way.

So is it all doom and gloom? Are you, as a manufacturing CEO, consigned to a life of misery, tied to a never-ending list of supply chain issues? Do you have to keep bouncing between possible candidates to take the manufacturing problem off your hands, or do you need to spend your future profits on external consultants?

The answer is *no*, and the solution is the basis for this book.

My Passion for Manufacturing

I started my love affair with manufacturing in 1987, at the age of 19, in the United Kingdom. Although I had worked in a wide range of part-time jobs since the age of 12, manufacturing was different. As a scholarship employee with British Gas I was immediately exposed to coal gasification sites and offshore gas rigs, where huge, powerful, and potentially lethal equipment, processes, and materials were all pulled together by skilled people. And it was the people that most fascinated me. It was as though those people were different every day. I never knew how they would respond or react to circumstances. The same instruction or experience could result in an entirely different behavior from one day to the next. That was the area in which I chose to build my expertise. How could the combination of people and processes deliver the maximum success for both the individuals and the business?

I graduated a few years later with a degree in chemical engineering and joined the world's largest consumer goods manufacturer, Proctor & Gamble (P&G), as a young manufacturing manager, working in Manchester, England. That was the start of a 17-year apprenticeship that took me to manufacturing facilities around the globe to which I owe a huge debt of gratitude. That 17-year period, which concluded with me being responsible for P&G contract manufacturing operations throughout Asia, gave me exposure to some of the best manufacturing facilities in the world.

In the ten years after leaving P&G, I spent four years as managing director for a manufacturing business with 4,000 employees and then became an advisor to manufacturing CEOs around the world. Overall, I've visited hundreds of manufacturing facilities in over 25 countries. In doing so, I have had the opportunity to not only witness but also test to find out what really works and to glean the very best practices. I have been able to rationalize and simplify until the golden nuggets have appeared, and then I have proven them in businesses that cross country and corporate cultures. I have tested the concepts across manufacturing industry sectors and with different generations of leaders.

Over the past 30 years, in roles from frontline employee to manufacturing CEO, I have made countless mistakes and learned from them. I still make mistakes sometimes, but I've learned how to minimize the possibility and impact of wrong decisions. I have polished the processes I present in this book in order to maximize your chances of success.

Although I don't know your specific company as well as you do, I have the benefit of having implemented the processes across a wide range of businesses and so I can give you a roadmap to follow. This book won't do the work for you, but it will be a step-by-step guide through the same processes that have worked for me and my corporate clients and that I teach in my online courses. The ALIGN Process will give you all you need to deliver rapid results and avoid the potholes.

How would it feel to have a manufacturing business you are proud of, and employees who are inspired by your leadership? How would it feel to become free from the daily issues and have the time to focus on business growth? This book will show you how to lead a successful manufacturing company in a way that is simple, rewarding, and more enjoyable.

As you complete the exercises given, you will establish the framework for transforming your manufacturing business, and you will know how to implement it in the most effective manner.

How would it feel to stop complaining and instead be able to say, "I give direction to the factory and it gets done right first time, without surprises!"

Full Disclosure

I have a secret intention. It has to do with your stress level and your potential to change lives for the better.

Over the past 30 years I have had the pleasure of working with tens of thousands of frontline manufacturing employees. Most of these people work extremely hard, often in some of the most challenging environments. In most cases, their families are dependent on their income, and their income is dependent on the success of the company. I have seen people rightly and wrongly hired and fired. I have seen people work 14 hour days with a two-hour commute. I have seen people laid off as a company or division folded, and I have seen new businesses rise from nothing, bringing hope to thousands. I have seen people's careers threatened by safety accidents, and I have seen companies fold due to poor output quality. And I feel for them. *In almost every case, the hardship to the employee and their family could have been avoided had a better process been in place.*

The book is a gateway to changing the life of the manufacturing leader and, therefore, the lives of the hundreds or thousands of employees that work for him or her. As the pressure on the CEO releases, because of clear direction and systems that allow him to help and support rather than instruct, the pressure also releases from the employees. People working for the CEO begin to take ownership for their areas of work and are given the help and support required to make smart decisions. In turn, they can better manage their own work and support the work of those around them. In this way, we unleash the estimated over 40% potential that is currently being withheld by most manufacturing employees, and target that energy toward the most important results.

When a manufacturing business is aligned, everyone succeeds. The customers, the employees, the shareholders, the suppliers, and the community all benefit. Most importantly, leadership becomes easier as the employees start to smile again and the cloud of negative pressure lifts. The employees become engaged and deliver, the equipment is restored to ideal condition, the facilities are beautiful and efficient, the products are the highest quality possible, costs are consistently good, and shipments are predictable and flexible to customer demand. At this point, the leadership team is no longer consumed with daily operation and can focus on innovation and growth.

It's the responsibility of the CEO and the manufacturing leaders to make it right – to make the leadership process work right and to make the products right. And it's time for you to get that sorted out. It's time to make your manufacturing facility efficient, clean, and productive, as you know it can be, and to make it a place where employees love to work and where customers want to have their products made. This transformation can be simple if you follow the steps in this book.

CHAPTER 1

Breaking the Mold

"The challenge is that the day before something is truly a breakthrough, it's a crazy idea. And crazy ideas are very risky to attempt."
Peter Diamandis

In picking up this book you have already demonstrated that you are open to change. Those who are willing to learn are the ones who are most likely to adapt. In the rapidly changing business environment, it is essential to adapt. Those who don't adapt will fail; those that do adapt have at least a fighting chance.

In this chapter, I explain how I waited too long before adapting; how I suffered personal failure before learning that there was a better way; and how I committed to learn from the very best in business so that I would not fail in the same way again.

From Breakdown to Breakthrough

I had been with P&G for five years and was flying. With two successful assignments already under my belt, I had moved to a new division that was making baby diapers at a factory in England. The process was a high-tech process and my role was challenging, but I was highly motivated. To fully support my team, I was working 12-hour days, spending most of my time on the production line, and really mucking in, hands-on. I was ensuring that all of the company systems were followed, was leading each product change myself, and was attending all the required meetings. I was doing everything I was supposed to do… and the results were terrible. No matter what I tried, I could not get the line to run consistently. Our goal was 85% efficiency and, for most of the first year I was there, we had been achieving 76%. My 12-hour days morphed to 14 and then 16 hours. And the results still didn't come.

After almost a year, one of the global directors visited the factory. That was a big deal for the three levels of management between me and him. But the plant looked like new, including my line. I had learned the importance of getting things looking great for a visit from a higher-up. When the time came for the director to visit my line, he looked briefly at the equipment, then at the results charts on the board, and said three words: "This is shit!"

I was stunned. I didn't hear much more except the clanging of my career collapsing around my ears. "This is shit!" meant *I* was shit. I had poured my life into that line for a year. I had tried all my best motivation techniques, worked closely with the team, kept smiling throughout, and fought all the issues that had come up. And I had failed, possibly for the first time in my life.

A few months before, I had bought my first house, on a 95% mortgage, and I slumped back to it that evening. Without really being aware, I walked up to the spare bedroom and sat on the bed. And then it came. I broke down. I started to shake uncontrollably, slid to the floor, and cried. I felt completely alone, useless, and utterly helpless. Perhaps most importantly, I'd lost control.

Within that moment I saw two clear options. One was to quit, to accept my failure, make things easy for the company, and move on; to take a break from the long hours, the struggle, and the stress; to give up. At least then I would be away from the sense of dread that going back to the factory filled me with. And I wouldn't have to face that damned situation anymore.

Then another choice swept over me and filled me with such a feeling of peace and calmness that it was almost spiritual. I can still feel it today. This second choice came from the realization that *I knew I was free*. I had tried everything that I was *supposed* to try and none of it had worked. I had done things the way I had been *instructed* to and *expected* to, and I had failed. Now I was free to try anything. Things couldn't get worse for me. Being fired would be no worse than quitting. I suddenly felt free to try things entirely my way. I could experiment. I could use this situation, this production line, as my guinea pig, and learn.

Within an hour I had redesigned everything about how I would run my production line, the teams, the processes, and the systems. I decided that I would work no more than eight hours per day. I would stop trying to do everything, including everyone else's jobs. I would let the team leaders make their own decisions for their teams and take accountability for their results. I would only be on the production line when requested by the team to help out or to verify the information I was being given. My role would be to fully support the requests of the team leaders, whatever it was they needed. Based on our common goals, the team leaders would set targets for their teams and report weekly on their progress and make requests for support for improvements. Also, we would reset the production line to its original condition, following good engineering principles.

That night I slept so deeply. Little did I know what was still to come!

The following day, I sprang up and went into work early to set about talking to my four team leaders. I laid out the challenge and what we needed to achieve, and I asked for their input. We outlined rough plans together, with the team leaders stating where they thought we had been going wrong and what they wanted to be doing differently. We set some guidelines, they chose some targets that they thought were achievable, and they promised to ask for help when needed. We put together a two-week plan to tear apart the production line, clean it, and rebuild it – to reset and start over. And then we got started.

After two weeks, the production line would not run at all. We were at 0% efficiency and zero product. With me sticking to my decision to limit my working time to eight hours and refusing to look over the shoulders of the people doing the work, there was a buzz around the factory that I had "checked out." The pressure was on, more than ever, but I remained calm. After four weeks of refusing to take shortcuts and following the strict engineering principles we had agreed on, the line was running at 73% efficiency. That was better than 0%, but worse than before the director had come to visit. But we stuck with it. I held off the executioners, who I'm sure were already looking for my replacement. I protected the team from all of the pressure and noise from above and took full responsibility for the situation. After all, I was the one with nothing to lose, but the team leaders had everything to lose. I knew the team leaders felt for me and I could see the effort they were putting in to make the line a success. For the first time, I felt really supported by the teams.

And then the breakthrough happened. The following month, we ran at 90% efficiency, which was a factory record. The next two months we also hit 90%, and we had the lowest ever scrap levels achieved by any line in the factory. And the teams were flying; they were achieving it all themselves. I was still working eight-hour days and focusing on simply supporting the operating team leaders and dealing with the pressure from above.

For the first time ever, I was out of managing the day-to-day details that had sucked so much of my time, energy, and focus. I was working strategically rather than tactically, doing the activities that would lead to sustainability and growth.

When it became clear that the new efficiency levels were not a fluke, I was promoted, to help the rest of the plant achieve the same results. Shortly after that, I was offered a position in Asia, the fastest growth region, to help develop the business there. I was on my way. And I had discovered a system that worked for different product lines and in different facilities.

———◇———

So, what did I learn that night I had the breakdown in my spare bedroom? The foundations included having a clear direction and strategy; having the conviction to follow through; allowing others to take ownership and be accountable for their own actions and results; setting aligned goals and regular reviews, with the intention to help; encouraging collaboration from the support departments and reducing pressure on the teams; maintaining the equipment in the ideal conditions; troubleshooting down to root causes; and building mutual trust and respect with teammates while eliminating hierarchical barriers.

As I progressed through different roles at different facilities, I learned more about how to apply these foundations in different businesses, cultures, and situations. As I studied and advised, helping hundreds of manufacturing companies, I honed the process and developed templates.

Most importantly, I learned how to keep the process simple, so that it was easily understood and could be rapidly rolled out to the entire organization.

Twenty-two years after that breakdown and breakthrough, having helped transform the lives of thousands of frontline decision makers, I still don't feel that I have done enough. This book represents my desire to spread the word and ensure that as many people as possible learn from my struggle without having to struggle themselves.

CHAPTER 2

The Manufacturing CEO

"… business is not that complex. It's just that there are too many people out there making it complex. The way to fight complexity is to use simplicity."
Jack Trout

Manufacturing contains all the traditional elements of business complexity: machines, systems, materials, logistics, and – the most complex of all – people, lots of people! Whenever people are involved there is a requirement for structure and discipline. Please don't assume that these words mean *restrictive* and *limiting*. When simply and effectively applied, structure and discipline can free the organization, giving rise to wonderful creativity and expansive thinking.

The story below illustrates an issue I see far too often in the many manufacturing companies I have visited. It demonstrates how quickly a situation can get out of control without the simple elements of structure and discipline in place.

The Contamination Challenge

You may well be familiar with this scenario. A manufacturing company spends millions of dollars on new equipment. Everyone in the organization is proud of the new equipment and wants to see it running well. They treat it with great care and spend time to set it up to the optimum conditions. And then they start to run it. Materials are processed, and materials cause contamination. When the equipment has not yet been optimized, there are a few material jams, which cause some mess, which is nicely cleaned up. The machine is started again. A few more jams and mess as the settings are being optimized for the specific materials. So far, so good. But then comes the pressure.

No one spends money on new equipment without a need for that equipment to be making product. And the product is required *now*. There is not time to keep cleaning and optimising. "Just get some product out, and we'll get the cleaning and optimising done when there's a break in the production schedule. In fact, let's plan to take the line down for cleaning next Tuesday, for four hours."

Have you been on that slippery slope? Where the equipment's condition is sacrificed for volume? Where the drive for the next product sets the priority for all other tasks?

The cleaning time on Tuesday comes, but because the machine was never optimized, the production volume was not made on schedule. All the additional equipment stoppages ate into the production time, and we are now behind. The four hours of cleaning are reduced to one hour. It's not perfect, but it's considered "good enough for now." After the insufficient cleaning time, the start-up of the line experiences some challenges, and that creates more contamination. "It would have been better if we had just kept running and not even stopped for cleaning!"

Within a month (or less, in the case of many companies) the expensive equipment no longer looks new. Within three months, the contamination has become permanent on some of the surfaces. Due to frequent material jams, the settings of the equipment have been changed to allow more variability in incoming materials. In some areas, the operators have made minor modifications with sticky tape and cardboard (you know this is not an exaggeration). The throughput of the line never quite met the performance targets promised by the equipment supplier, and it seems to be getting worse.

And now you are lost. You may not know it yet, because the machine is still producing. You may even have a few record days of production as the operators figure out, through experience, how to modify the line settings to their particular standards. And now you have the added variability of operator "experience."

I've just realized I'm holding my breath as I'm typing. Because this stress I'm describing is present in almost every manufacturing facility I've visited. It is so endemic that when people ask me what I do in my business, I end up telling them I spend 90% of my time encouraging manufacturers to keep their equipment clean and properly aligned.

I've seen the situation I've described above time and time again when the priorities and the strategy for the business do not effectively support growth. When, on the other hand, a more supportive strategy is reinforced with structure and discipline, the foundations for a successful manufacturing business are set, and problems like the ones I described occur only rarely, if ever.

The Frontline Gap

"A unique strength of Mr. Snook is his ability to bridge the gap between the line operators and the management team. Kevin can talk to them all and ensure alignment. I have rarely seen this capability in consultants who normally want to set strategic direction and leave it for others to implement"
B. M. Choi, CEO, KleanNara Company, Limited

In every manufacturing company I have observed, there is a visible gap in communication, direction, and expectations between the frontline operating teams and the management. Where this gap is largest, the company struggles the most. Once this gap is closed, however, success follows. This is a people gap!

It sounds simple, and I believe it is. *The issue is that most companies either don't recognize the gap or don't have a process to close it.* In many companies, "people" are considered the responsibility of the human resources or personnel teams. In companies that are slightly more enlightened, people are the responsibility of the direct line management, but in many cases the management are not effectively trained to deal with multi-faceted human behavior.

The challenge with people is that they wake up with a different head on their shoulders each day. Swimming around in that head are numerous different personalities, different characters, and different stances. An opinion one day can be reversed 180 degrees the following day, depending on the circumstances. This is the basis of both the joy and the pain of managing people.

Take the following true story of Jim, an operations team leader on B shift. Married with three young kids, Jim has always been a model employee. He quite likes his job, has risen to the role of team leader, and is well respected by his team members. Although his boss is young, much younger than him, he gets on with his boss quite well. Because Jim already knows the job inside and out, there is not really much of a challenge on a daily basis. Recently, though, he has been having a few challenges at home. His wife has been going out more often in the evenings and, as he has been tired from his shifts, he has been putting the kids to bed and then sleeping himself. Although he wants to ensure that his wife is still happy and having a life, he questions whether she's really spending all that time with her friend Carol. She didn't seem to want to spend much time with Carol during their last discussion on the subject, but now they're apparently best friends again, going out together two or three times a week and staying out late. Is his wife really out with Carol? Is he being played for a fool? Is there something deeper going on, something that he would never have considered only a couple of years ago? Aren't he and Carol super happy together? After all, there was such a connection when they first met. They're soul mates! Aren't they?

It's Thursday morning and Jim didn't sleep so well last night. His wife got home after midnight and, prior to that, he was finding it hard to sleep. After she returned she had a shower and climbed into bed without a word. Wasn't that strange? Then again, Jim was pretending to sleep. Why would she say anything and wake him up? Or was she hiding something and preferred to stay quiet? Didn't they always used to cuddle before and during sleep? Now they seem to spend more time with their backs turned toward each other.

As Jim arrives for his shift at 6:45 a.m., he sees the look on the face of the outgoing team leader. Clearly, something didn't go well last night. He soon finds out that the size change on the production line hit all sorts of problems and that the line hasn't run for the past six hours. It's going to be up to his team to get the issue fixed and get the line started up again. This is the biggest challenge in an operation like theirs, a challenge Jim would normally relish and rise to. But then he remembers that today has a critical deadline. This is the day that they want everything running smoothly, because it's a huge day for Jim's direct boss. The line has to perform at its peak to meet the target. Jim knows he could make it happen. Suddenly, a flashback enters Jim's mind. Prior to getting married, he once saw his wife chatting to a young and quite handsome guy in the pub. One of the players from the local rugby team, maybe? Didn't his wife seem to return to the table that night with a glow around her similar to the glow that she's seemed to have over the past few months, ever since she started going out at night more often?

Jim hears a voice that comes from behind him, but he doesn't really hear the words. He turns to see his boss, Peter, waiting for an answer.

"Sorry. What?" Jim says.

His boss, slightly irritated, asks snappily, again, what happened during the night shift and what Jim will do today to get it all sorted out? He goes on to ask why Jim is so spaced out. Doesn't he realize that today is a big day and the line needs to be running? What's the plan?

Jim says that he has no idea. He asks why it's always up to him to sort out the mess. It's the idiots on the night shift who are continually causing these kinds of issues. And what about his boss, Peter? What was *he* going to do to get it all sorted? What did Peter ever do to help the team? For years, Jim had been asking for more training for the newer team members, for the right tools to get the job done, and for a specialist to help with the most challenging equipment. And what had they been given each time they asked for help? Nothing. That's the way it was around here, one excuse after another, but no action. The management had deaf ears!

"This is the one day I expected more from you," replies Peter. "It seems you really don't care about all this. Just go home. I'll ask Gary to stay on overtime to get this sorted. But when you get back tomorrow, make sure that attitude of yours has been sorted out."

So, what had happened there? Was it a perfect storm of mishaps? No. Not really. These types of internal challenges are going on in our minds and the minds of those around us every day, to some extent. And we are expected to continue achieving world-class results despite that noise. While confrontations between people in each department, level, and team are inevitable at times, how do we build a structure that allows conflict to be resolved in a manner that always supports the company's direction? How do we ensure that, even with the inevitable differences between individuals, we eliminate the gap between what is expected and what is delivered?

Since every decision in every company in the world is made by people, it is the role of leadership to understand and support their people. Closing this frontline gap is the key to the organization delivering on expectations.

The Expectations of Manufacturing CEOs

There is no surprise that the role of the manufacturing CEO is considered to be one of the most challenging jobs. While a large amount of the complexity and variability can be predicted using certain models, humans make all of the decisions in the process. These decisions are being made every second of every day, so the key is to focus on people leadership. The role of the manufacturing CEO is not to manage a supply chain, a factory, or a production line. It is to inspire a vastly diverse group of individuals to make the most effective decisions every second of the day, no matter what thoughts they wake up with.

The Organization Pyramid

Most manufacturing organizations are working under a huge misconception – the idea that the CEO sits at the top of the organizational pyramid, with the frontline workers at the bottom. In this outdated model left over from the times of the Industrial Revolution, instructions are passed down the levels of hierarchy from the top to the lower levels, where the workers implement the wishes of the bosses. When the workers hit an issue, they pass information up the levels and wait for a decision to be made and instructions to be passed back down to the workers.

When tasks are easily separated and instructions are crystal clear, this system works relatively well, despite the time lag in getting decisions made. But that is not how the world works today. With the need for higher productivity, each employee is being asked to do more tasks with higher complexity. They are required to make decisions, not only implement instructions, and this creates a conflict. When is a frontline employee expected to make a decision and when is she expected to ask for a decision to be made? Asking for a decision can take hours, days, or even weeks. Meeting after meeting might be required for an answer to be given. On the other hand, making her own decision and getting it wrong can result in losses for the company and could lead to dismissal for the employee. So, what to do?

In my experience, most company leaders would agree that the frontline employees do need to have some authority to make decisions. In fact, those employees are currently the ones making the majority of decisions for the business, 24 hours per day. While managers and leaders may think they're in positions of authority, they really have a limited impact on the day-to-day decisions made by the frontline employees. Who decides whether or not to clean the machine well, to shut down the line to fix a quality issue, or whether to tighten the drive belt to the right torque?

With the frontline employees already making the vast majority of decisions, and with the lack of capacity or ability of the leaders to make those decisions, there's a logical solution: *Help frontline employees make the best decisions they possibly can.* The employees need to make decisions in a manner that takes into account the entire business, not only their area or division. They must make decisions that drive the overall business results forward, despite possible impacts on their personal or divisional results.

But how, and why on earth, would an employee sacrifice her own personal results for the sake of the business? Would anyone really do that?

The answer is *yes*, but in order to tap into that state in an employee, there needs to be a new type of revolution to supersede the Industrial Revolution's assumptions about frontline workers. This requires a revolution in understanding, and that starts with the role of the leaders.

There is one major expectation of the CEO, and that is to make the organization successful. Whatever metrics are used to measure that success, usually growth is the mandate of the CEO. *The best way for the organization to be successful is for every employee in that organization to be successful.* In a well-aligned business, for the employee to be successful he must deliver results that move the organization forward.

Notice that I say "in a well-aligned business." In many organizations, it is possible for individuals to get ahead at the expense of the organization. Imagine a production worker who purposefully makes off-quality product in order to boost the volume numbers at the expense of the customer. At 3 a.m. during the night shift, with the quality inspector on a break, it is much easier to let a few defects slip through to boost volume, rather than make the difficult choice to stop the line, fix the problem, and face the wrath of the boss in the morning for missing the production target. The employee who makes the easy choice may be considered high-performing, especially if the quality monitoring system is inadequate and does not detect the defects. In the short term, and sometimes even in the medium term, that employee could be rewarded with bonuses or a promotion, all at the longer-term expense of the company.

For the CEO to be successful and to ensure that every employee is making the right decisions to drive forward the overall success of the business, there are several elements that need to be in place.

- First, the CEO must set a clear and compelling direction for the business. For all employees to row in the same direction, the destination must be both absolutely clear and somewhere that everyone wants to get to. To reduce the need for constant external motivation, this goal ideally needs to be inspirational to each individual, so that they are self-driven to get there.

- Second, the role of each employee must be absolutely clear. This allows ownership, accountability, and a direct understanding of personal impact on the result. Along with providing opportunities to make decisions that directly affect performance, this ownership is a critical element in making the individual's role worthwhile to them.

- Third, there must be immediate feedback on the results of decisions made. Bowling without being able to see how many pins have been knocked over is not only no fun, it makes it impossible to measure improvement. Feedback is a critical element of continual improvement, and immediate feedback accelerates progress.

- Fourth, there must be an opportunity to ask for help and support when it's required. When an employee needs to make a decision outside of their area of influence they need to be supported. Requests for specialist support, training, and guidance need to be rapidly fulfilled.

- Fifth, there must be recognition. However inspired an individual is internally, there is a desire for rewards and recognition. Milestones achieved deserve to be celebrated.

Individual recognition is valued and can make an impact for a lifetime, if well done, but team celebration has a more widespread benefit. Both are important.

These are the elements of the process described in this book. As we work through the process, these elements and the expectations of the manufacturing CEO will be clarified, with specific deliverables identified.

Moving back to the issue of organization pyramid, in my experience, turning the organization on its head, putting the CEO at the bottom in the role of supporter, is more in line with current requirements around how manufacturing businesses actually work. The frontline employees, now at the top of the inverted pyramid, are the ones making the majority of decisions. They are the ones directly in touch with the product and, therefore, the consumer. When making decisions, there are times frontline employees need to ask for help and support, and it's the role of the leaders who are further down the inverted pyramid to rapidly provide that support. When the CEO adjusts the direction of the company, the entire organization, led by the CEO, responds. When the CEO develops systems, expectations, and structure, the entire organization is supported.

For the CEO to be successful in optimizing the business results, *every individual must be successful in their area of ownership*. The role of the CEO, therefore, becomes giving every employee the direction and support they need in order to be most successful. But with individuals being so different, how do we know what they each need? We listen with the sole purpose of giving help and support. This becomes a critical requirement for every leader in the business.

The Leader's Ego

Ego is a funny thing. It masquerades as a supporter or a defense mechanism, but it actually prevents success. Whenever we are massaging our own ego, we are sacrificing service. In order to ensure that every employee is successful, our full-time job as leaders is to be of service.

Have you ever stood in front of a room full of people to give a speech and, before you went on stage, you worried about your hair, whether your tie was straight, and whether you had prepared the right speech? All of these factors are your ego speaking. They are all issues about you. Yes, even the thought of whether you prepared the right speech for the organization is actually all about you. It's about whether you will look okay and whether you will succeed.

There is an alternative. When you are in the mode of service, it is no longer about you. While giving your speech, you may be asked a question. Rather than referring to your script, you could give an open, honest, and vulnerable answer. You can step outside of what you think you *should* say and step into what you feel you *want* to say. At that point, the angle of your tie no longer matters. All that matters is giving the individual who asked the question an authentic answer.

We fluctuate between a state of service and a state of ego. The strange thing is that many people spend far more time in the difficult state of ego. In an egoist state everything is calculated. Everything is planned. Everything is controlled. And this state is transparent to others. We are all designed with inbuilt lie detectors, and the alarms go off when we are fed an egoist response. Credibility and trust are lost. Some people have become quite expert at faking a state of service, but this is exhausting. We are always concerned about being found out, about having our ignorance revealed, but when we choose to focus on being of service, those concerns don't matter so much.

The state of being of service is absolutely easy. It simply is what is. When you don't know something, you say you don't know. When you can help, you say you can help. When you won't help, you say you won't help, and you say why. When you are uncomfortable, you say you are uncomfortable. When you are angry, you say you are angry. And all along you keep service in mind. How can I best serve? How can I help everyone here be successful, so that the organization is successful and we move in the desired direction?

This mode of leadership has been called many things, none of which really hit the spot for me: servant leadership, vulnerability, authentic leadership. Though they are all correct, the main point to focus on is the very simple requirement of putting your ego last and service first.

The Culture Excuse

"That would be difficult here, because we are in…" Korea, Argentina, Indonesia, China, Germany, or wherever else.

"You don't understand. That's hard to change, because we've been in business for…" 120 years, 50 years, 10 years, 2 years, or however long.

"That sounds great, but our employees are…" too new, too experienced, to old, too young, skilled engineers, unskilled labor, contract workers, union affiliated, or whatever else.

I've heard every culture excuse – about societal culture or business culture – under the sun, and the only common theme is that they are all excuses. This is a hard thing to accept, because we all like a good reason to explain away our less than perfect performances. The bottom line is that the CEO sets the culture for the entire organization, and the individual leader sets the culture for his or her group. There's nowhere to hide!

People do not leave organizations; they leave individual leaders. People do not work for organizations; they work for individual leaders. You either have leaders in place who are setting the "right" business culture, or you have leaders who think of themselves as victims. The good thing about this is that you have a chance to choose and set the culture you want. The challenge is that you have to know the culture you want. If in doubt, go for a culture of success for each individual, because that leads to the success (growth) of the business, with transparency as the operating mode. And the role of the leader is to set the direction and then help and support all to move in that direction.

The Time for Change

So, you are already working 16-hour days. How are you supposed to find the time to implement a new strategy, to change the culture, and to build the capability of the employees? And time is only one of the factors you manage. You simply don't feel like you have the energy to take on such a big rebuilding project.

Well, there is a secret to this. In my estimation, from studying hundreds of managers, I believe that approximately 40% of what you are currently doing adds no value to the business. In some cases, I would increase this to 60% – and not only for the CEO's productivity, but for almost every leader in the business.

How can this be possible? Let's imagine a scenario. You have found out that you have a particularly nasty strain of cancer. For the next three months you are going to be in chemo, radiotherapy, and going through other treatments for 50% of your time. During the other 50%, you will be required to be in recovery most of that time, rebuilding and maintaining your strength. All of that will allow you one hour and 20 minutes per day to do your daily work. What would you do in that one hour and 20 minutes per day?

My guess is that you would spend ten minutes each talking with your key team members, to see if they need any help or guidance and to give them some encouragement. Perhaps then you would delegate some tasks, then talk to an important supplier or customer to see if they need any help and to reassure them. You may even need to sign a few critical papers. Let's assume that, during this time, one of your direct reports really did need an hour of your time to work through an issue and get your support. My guess is that would be your priority for that day, and so some of the signing and other updates and check-ins would need to wait.

After three months of handling things this way, how much would the business have suffered? *Do you really believe you are so important that you need to spend ten hours per day doing other people's work for them?* Once you've recovered and you've gone back to work, how long do you think it would it be before the meeting times had extended, the work had piled back up on your desk, and you were being asked to make decisions that, until recently, had been successfully made by others?

There is always time to make a change. It is your job to make changes for the better, to challenge the status quo, to strive for growth, and to build capability. If you don't have time for that, you are mismanaging the company. Yes, you will need to stop doing some of the things you are doing now. Yes, you really must.

The Solution

Let's assume, because you are reading this book, that you are the type of leader who will drive forward and take your team along for the experience of the ride. This book is designed to simplify that process; to take you step by step along the journey; to clarify the direction, align the team, create the plan; and to put the plan into action. It distills the best practices I have experience with down to the ones that are proven to work. It shows how to avoid the pitfalls and complete wastes of time that many people still advocate. It brings reality and simplicity to a process that the Big Four consulting companies would love you to think is not possible. But I know from repeated experience that it is.

The ALIGN Process

I call the five steps I'll lead you through the ALIGN Process. You can use it to ensure that you lead your manufacturing company in a way that is simple, rewarding, and highly successful.

Follow the steps. Each step is the answer to a question I often hear from CEOs. Follow the Action Items for each step, which will guide your decisions.

Don't be fooled by the simplicity of the steps. Simplicity is essential for helping your team fully grasp and align with your mandate.

These are the five steps of the ALIGN Process:

1. A – Aim from the Heart

2. L – Lead with the Frontline

3. I – Inspire with Information

4. G – Give Help and Support

5. N – Nurture Feedback and Recognition

Each step is presented in more detail in the five chapters that follow. Each chapter starts with an introduction to the step and then helps you develop the answer to the most frequent questions that challenge manufacturing CEOs. The main questions presented for each step are in the form of what you might currently ask as you're first reacting to a problem. With the aid of this book, I hope to encourage you to pause for reflection at any sign of a problem in your manufacturing business. Within that pause, I encourage you to switch from acting on an immediate reaction to having a more measured response that you act on in a manner that most effectively aligns the organization with growth and stability.

In the next chapters, action items are clearly identified, along with relevant exercises to support you through each process.

For each step, there is an example that can serve as a reference, along with bonus materials for further guidance.

Throughout the presentation of the five steps, I'll use *vacation planning* as a simple example, one that you can also use when guiding your organization through this process.

---o---

CHAPTER 3

---o---

A: Aim from the Heart

Does "Aim from the Heart" sound a bit too touchy-feely for you? Do you feel more like the manufacturing organization needs a kick up the backside? Believe me, I get it. The complexity discussed above creates a level of stress and urgency in a manufacturing company that is quite unique. Bear with me, though, and I'll explain in depth why the desire to "jolt" your organization into producing more volume may be the exact behavior that is holding your business back.

This chapter explains how and why it is critical that you:

- Set *your* destination and direction for your organization.
- Set the framework of the action plan.
- Set the ground rules and expectations your team will follow.

Step A of the ALIGN Process helps you set these factors in a manner that compels, excites, and enrolls the organization.

Due to the foundational importance of this first step, I have broken the discussion down into three reactive questions often heard by CEOs from people within their organizations:

- "What were you trying to achieve?"
- "Who is accountable for that?"
- "How could you let that happen?"

Let's start with the first of the three questions.

What Were You Trying to Achieve?

"If you do not change direction, you may end up where you are heading."
Lao Tzu

It's impossible to start an effective journey without a clear destination.

When planning a vacation you may be able to figure out some directions on the way, especially if you are going alone, but if you have an extended family in tow the situations and pressures are seriously multiplied!

For a successful journey everyone should know where you're going and, just as importantly, must want to get there. Going to a distant relative's house may not generate the same level of commitment to the cause as going for a beach vacation in the Caribbean. When the going gets tough, as it inevitably will, it's the shared direction and desire that pulls the group through, allows the entire group to cheer their progress, and makes the trip successful.

Within the main question, heard as a reaction from the CEO ("What were you trying to achieve?"), there are others in the minds of the employees:

- "Where is the organization going?"
- "What are we aiming for?"
- "What is the purpose of all the hard work?"
- "Who will benefit?"
- "How will we know when we get there?"

Although employees may not ask these questions regularly, they do think about them. When the organization's purpose is clear and compelling, when it generates shared excitement and commitment, then that gives every employee a reason for working that's in addition to "making some money to support my family." Although most people don't like being told what to do, they do like to know where their efforts are taking them.

The Two Roles of the Leader

Look at a thousand sources, and you'll get a thousand answers to the question, "What is the role of the leader?" For me, the answer has become increasingly simple the longer I've worked in business. There are two primary roles of the leader:

1. To set direction
2. To help people get there

I feel this issue is often overcomplicated. We talk about motivating, inspiring, setting rules, building the culture, dealing with outside influences, preparing contingency plans, and on and on. But what we all want from a leader is for them to tell us where we are headed. We don't even need to know the final destination; in fact, in business there is no final destination. We just need to know where we are heading and if we are making progress. Then we occasionally need help to get there. We need help to be successful in our own rights in order to help the organization be successful. If we are all aligned, then individual successes will lead to the business's success.

The First Role: Set Direction

Let's make this as easy as it can be. Where are we going? In which direction are we headed? And how can this be so clear that we are all aligned in our efforts to get there?

You are the leader. You *must* have a clear direction in which to take the organization. A good direction is compelling for you. It is somewhere that you will be willing to be a hero to reach, and that motivates you to the point that no matter how many times you get knocked down you will get back on your feet and drive forward. You will be entirely focused on moving in this clear direction. This is the primary role of the leader.

So what is your compelling direction? What is a need that's so much greater than yourself that you are willing to get up every day and strive toward it? A need so compelling that you simply *will* take steps forward every day. A need that you will keep yourself fit for, focused on, and dedicated to achieve. Where are you going and why? Write your thoughts down now, and we will fine-tune your direction throughout this step.

It's Your Direction

Your direction is going to feel extremely personal. After all, it is *your* direction. It is not necessarily the direction the business owners told you to take the business in. It is not necessarily what you think your employees want from you. It is, most likely, not what the board of directors expect. But they are not the leaders – you are. This is *your* compelling direction *and* the direction you will take the business in.

Isn't this risky? Absolutely. Apple fired Steve Jobs from his own company for taking this approach. Elon Musk was close to extinction as a business leader for this approach. Now both are heroes. So, let's get real. The risk of following this approach is that you will find your passion, your compelling direction for the business, and get fired. Sometimes that happens to heroes. And they get back up, dust themselves off, and go again in the direction that is so compelling for them that they will eventually succeed. Once you start down this path it is very hard to go back, to compromise. No one is stopping you from going back, but the future becomes so clear that there is no desire to go back. You will be compelled to move forward. The masks will fall away, and your authentic self will be revealed. And then you become an inspiration to others. Then you attract followers.

Inspire

There is a distinct difference between motivation and inspiration. Some call this a difference between external and internal motivation, but I'm going to stick with motivation and inspiration. In my simplistic terms, motivation is a force exerted on your mind from outside; inspiration is a force exerted from your heart to the outside. Let me explain a little more.

A child is the perfect clay for molding and for observing. Watch a very young child who is being encouraged by parents to reach for an object. First, there is a mental reaction of *What do they want me to do?* Then there is effort. The effort will last for a short period. And then there is frustration. Action stops and distraction comes into play. At that point, additional encouragement is given, which brings the child's mind back to the task at hand, and there is another short period of striving to reach the object. This cycle is repeated until the child reaches the goal of getting the object or their parent gives in and hands the object to them. In its best form, motivation results in small, short bursts of effort toward achieving a task, guided by encouragement. Without encouragement, there is no continued effort.

Now watch a child when they are alone. There is a heart-driven instinct to reach an object. With no external encouragement, no desire to please, no outside manipulation, the child will work, over and over again, to achieve the goal. Driven by a force inside, an internal desire to progress in a certain direction, the child will show incredible levels of dedication to moving forward. Whatever obstacle is in the way, they will find a way around it. There will also be incredible frustration, and this drives further internal determination. Often, when a young child reaches for the object of their inner desire, they will exert so much effort they fall asleep mid-process rather than turn away without the achievement. It is through our heart-driven desires that we progress most rapidly.

If you haven't had an opportunity to watch a young child in action, you can think of the most dedicated sports stars, actors, or business people. We often call what they have charisma, star power, stage presence, or an aura. An incredible energy emanates from such people. It comes from their incessant drive to move in a compelling direction, despite the obstacles put in the way. These people are not mailing it in. They are not "pretending" to get the job done. They are not wearing a mask to themselves. They are compelled to move forward by an inner drive. And they attract us just by being themselves. These are the modern-day heroes. To lead a world-class business, you need to become a hero.

Your compelling direction for the business must be inspirational *to you*. In order to generate dedicated, passionate, inspired followers, you must yourself be an inspiration who people want to follow – an inspired leader!

This is your first and possibly your largest challenge. In a business that makes widgets for profit, where is your inspiration? Where is your compelling need? What is your "big why" for getting up every morning and driving forward?

If this is something you're struggling with, then you are not alone. If, like many of us, you have become shrouded with indifference, hypnotized by mediocrity, and distracted to the point of despair, you are with 95% of the people I used to meet on a daily basis. Now I choose to surround myself with different influences and influencers, because I made different choices. You can choose to be different.

Create Followers

When leading a team, of course you need to know where you are all going, how you will get there, and how to do your best to enjoy the journey. But how do you get people to follow?

In a highly regarded global study (Kouzes & Posner, 2007) of what attributes a leader must have for people to follow them *willingly*, there were four attributes most often chosen by respondents:

- Honest
- Forward-looking
- Inspiring
- Competent

The attribute that stands out to me, and that was chosen by 69% of the study's participants, was that the leader is inspiring. I'm drawn to this attribute because, in my experience, many leaders consider themselves to be honest, forward-looking, and competent, but far fewer leaders have described themselves to me as feeling like they inspire their organizations. We discussed above the difference between motivation and inspiration; if people are inspired, they don't have as much need to be motivated, because there is an inner drive to move forward in the set direction; there is an inner fire burning to support the team, do the right thing, overcome barriers, and do whatever it takes to move forward. Imagine a team of inspired individuals pushing your business forward.

As leaders, how do we inspire? One way to look at this is to consider who inspired you. Who was a leader that you wanted to follow? Who would you have climbed mountains for, swum oceans for, woken up driven to serve? And what were their characteristics?

In my talks with young professionals, I often watch them carefully as they enter the room. What is their demeanor? Do they look like they want to be there? Is there a spark, a purpose? In certain individuals, there is an aura. It's not about being bubbly or being serious, making eye contact, being well-dressed or approachable, sitting at the front, being fit and healthy, etc., although those are some of the more common traits of people who are successful. It is about being there to learn, to absorb knowledge, to get every edge or advantage possible to take the next step forward. The inspired individuals demonstrate an inner drive and intention to move and change in order to achieve.

There are seven billion different personalities in the world. *I have never seen a particular personality type be a requirement for success,* but I have seen that inspiring others to move forward requires an obsession to learn and change.

The Second Role: Help the Individuals Succeed

The first, and most important, role of the leader is to set direction, and the second is to help the individuals succeed. Assuming that the employee and the leader (company) share a direction, then making sure the individual is successful is the primary method to ensure that the company succeeds. Alignment about direction is critical, and we will discuss later the importance of effectively communicating the direction.

Making sure an individual is successful can take different forms. For example, if the leader of the accounting division's compelling direction is only to be a professional photographer, then it is unlikely (but not impossible) that helping the accountant reach her photography goals will help the business. However, should the accountant desire to run her own accounting company, then there is absolute synergy. Why not help the accountant and her team develop the skills and capabilities required to spin off into a profitable, company-owned division?

In each onboarding discussion, the first thing I mention to employees is that my job is to help them become successful, whatever that means to them. For those who have been well recruited and have a joint compelling direction with the business, this is easy. Success for them in the company is success for the business. For those employees who have a different compelling direction, then it is a matter of seeing how we can best align the two directions. I am a realist, and I know that the majority of people do not currently come to work dreaming only of the business being successful, even though a successful business should benefit all employees and stakeholders. People have their own hopes and dreams that can be attained in conjunction with being aligned and successful in the business. In fact, this is my preferred type of employee – one who has an aligned, compelling desire inside the business, and another outside of the business.

I suggest to those people who are disinterested in the direction of the business that they find somewhere else to spend such a large part of their life. In a world of infinite opportunity and infinite competition, a world-class organization needs a fully committed team. It's not fair to any individual to keep them in a business in which they have no interest. Let them go do something more engaging for them.

Work tirelessly for those who choose to be an active participant in your direction to help and support them to be successful, in all their endeavors inside and out of the company. Like a young child reaching for the object of their inner desire, an inspired leader will exert so much effort they fall asleep mid-process rather than turn around without achievement.

Now let's take a look at the second of the questions heard from manufacturing CEOs about this step.

Who Is Accountable for That?

"One of the problems with posing a 'bold new plan' is that you can't just extrapolate from previous plans."
Nathan Myhrvold

Let's take another look at the vacation example. We know we want to go to the Caribbean and that there will be up to 20 travelers in the group. Now we need to know roughly how we will get there, what the major forms of transport choices are, the dates for leaving and returning, what the weather may be like, and whether we will spend time on the beach or in the city or both. Are we likely to play golf, go scuba diving, or do deep-sea fishing? Will we plan to camp, stay in a hostel, or stay at a five star resort? And who is going to handle making decisions around each of those issues?

Having a framework allows us to choose the right people and resources to fill in the details of the plan. A framework limits the infinite options and sets the tone for the experience. It allows the travelers to know specifically if they are interested in the journey and the destination and are willing to spend their time, effort, and money to get there.

In your organization, it's important to set a framework so that the entire organization understands the structure of the journey.

Within the main question of "Who is accountable for that?" there are others:

- "What are the framework decisions that are already made for the organization?"
- "What's your intention about how the business moves forward?"
- "How do you think you will most likely get there?"
- "How will you structure the organization?"
- "What groups will be needed at each stage of the journey?"
- "What will be the main roles and responsibilities of each group?"
- "Where will the boundaries between departments be?"
- "What will be the operating strategy that guides each decision?"
- "Who will be hired and how will they be hired?"

The content of this section of the book will guide you to answers for your organization and ensure that "who does what" is absolutely clear.

Everyone Wants the Same Thing

One of the most common complaints I hear from all levels of manufacturing organizations is that the other levels don't support or don't align with this level's priorities and the actions they want to take. The following story shows a different perspective.

I was recently in a meeting with a group of production line operators and asked the following question: "What would you most like to change about the organization?" I allowed people to work in small teams and then present their answers to the whole group. The teams then ranked the answers, which were:

- Equipment to be maintained in better condition
- Opportunity to make more decisions ourselves
- Better quality of incoming raw materials from suppliers

- Better physical working environment (workplace conditions)
- The minds of the leadership/management

I thought that was a great list. If we could significantly improve those items, the business would certainly progress. It showed a desire from the operators for improvement, and some rationale.

That prompted me to ask the same question to the next level of leadership, the direct managers of the frontline operators: "What would you most like to change about the organization?" Without knowing the answers that had been given by the production line operators, they came up with the following list:

- Equipment to be maintained better
- Operators to make more of the decisions
- Better raw materials
- Operators to follow the standard operating procedures
- The minds of the operators

Intrigued by how similar those lists were, I went on to ask the next level of leadership the same question. Their answers were:

- Maintain the equipment condition more effectively
- The minds of the operators to take more initiative
- People to follow the work rules and standard operating procedures
- Improve the quality so no risks

Those answers were not items picked from a list of choices. They were original suggestions made by groups that were independent of each other. As I looked at those lists, something became alarmingly clear.

All groups wanted the same things, and all groups felt that there was resistance from the other groups to achieving it!

What was also clear was that at the leadership level there was a contradiction between a) wanting the operators to take more initiative and make more decisions, and b) wanting operators to follow the rules and a set way of doing things.

There is a constant conflict going on in the minds of the leaders between allowing creativity and wanting control – doing things consistently to get the same result or doing things differently to get a different result.

The key to resolving this is *clarity*. When the direction of the company is so clear and compelling that people want to achieve it, the next step is aligning the actions. This involves making sure that everyone has an opportunity to contribute, and that the combined energy of all individuals is released and finely channeled toward the actions that will combine to give the desired result. This is the purpose of action planning.

Structure is Freedom

Structure is considered to some people to restrict creativity, to restrict freedom. If I am the kind of person who needs to color outside the lines, then it might be better to give me a blank sheet of paper with no lines at all. Creativity can release boundless energy, but it is important that the energy released takes us in our chosen collective direction. One individual's race for freedom may jeopardize the entire mission. So how do we ensure that individual desires to do what is right are aligned with what will work best for the entire organization? The answer comes in the form of an action plan.

The Action Plan

The action plan is the necessary structure to ensure that the organization moves forward in an aligned and effective manner. This action plan clarifies the specific roles and responsibilities of the collaborating departments.

No Olympic rowing crew has ever won the gold without structure that the entire group in on board with. Who is going to do the rowing? Who will steer? Who will prepare the right foods to promote health and strength? Who leads the training sessions? Who sets the alarm clocks and makes sure everyone shows up on time? Who maintains the boat? To win the gold, every department must be playing its part to the highest level. One loose bolt in the seat of the boat and the race is lost. One wrong chemical in the food and the team is disqualified.

An effective and aligned structure does not restrict the energy of each team member; it directs their energy to the places it will be most effective for the overall direction and goals of the team.

The Core Group

In the rowing boat, it would be hard to deny that the rowers are the core group. It is this group that all others on the team are designed to support. Is it fair to call out a distinction between the groups, even when one mistake by the boat maintenance crew could cause failure? Well, fair or not, it is what it is. A boat maintenance expert is not a rower, and a rower is not a boat maintenance expert. They both have equally important roles and aspirations; they just happen to be different.

In a manufacturing company there are two core roles:

1. The people who actually touch the equipment to make the product, without which there would be nothing to sell (manufacturing operators)

2. The people who contact the customers to sell the product, without which there would be no money coming into the organization (salespeople)

I want to be clear that this does not include the leaders and managers of those departments. Those people are in support roles, as is everyone else in the company.

So why do I distinguish those core groups? Because it shows the simplest level of structure needed to clarify whom we serve.

- The manufacturing employees serve the customer by producing the best-quality, best-value product on time.
- The sales people serve the customer by meeting a need the customer has for having the best-quality, best-value product when they need it.

Everyone else in the organization is there to ensure that those teams can do their jobs in the most effective manner.

Does this mean the core groups are the most important? No. Without all of the other groups playing their equally important roles there will be failure. However, now that we know whom we serve, we can make sure we meet their needs.

The Purpose of Each Department

There are many departments in a manufacturing organization and each organization seems to choose, often based on organic growth, the boundaries of those departments. You can, of course, design the department for yourself based on your company's situation. What is not negotiable is to have clear boundaries about departments and roles. This is not an endorsement of operating in silos or lack of collaboration, but defined boundaries are essential.

Here's a list of some typical departments and what I see as their strategic purposes:

- Production – To make product that meets quality specifications and the planned timeline in the most cost-efficient manner
- Technical Support – To ensure that the operators have the skills and equipment required for maximizing output
- Planning – To ensure that decisions are made to support meeting the delivery needs of customers while minimizing inventory
- Plant Engineering – To ensure that the manufacturing facility is maintained in "as new" condition, with the lowest overheads and utilities costs
- Warehouse – To store required inventory and ensure that it is well-maintained, tracked, and rotated
- Quality Control – To ensure that materials and finished goods meet the specifications and that no off-specification product reaches the consumers
- Quality Assurance – To ensure that all aspects of the facility, the people involved, the systems, the machines, and the materials support the quality goals of the company
- Human Resources – To ensure that the plant has a pool of talented people to recruit from and that the labour laws are complied with
- Information Technology (IT) – To ensure that people have the data and systems required to complete their tasks in the most efficient manner
- Plant Finance – To track all spending, help prepare return on investment (ROI) reporting, and partner with employees to make effective cost decisions

- Purchasing – To ensure that the materials, consumables, and machine parts are procured in the most cost-effective and timely manner to meet the needs of the business
- Payroll – To ensure that everyone gets paid correctly
- Research and development (R&D) – To ensure that consumer needs, as expressed by the marketing department, are translated into the best possible products and specifications, that can be manufactured effectively
- Marketing – To assess the needs and desires of the consumer and ensure that the research and development (R&D) department understands these needs, and to present product features to consumers and customers in a manner that drives demand
- Sales – To help consumers and customers find the product that fits their needs, and to take orders for those products

You can see that every one of those descriptions is simple and there is no overlap. I am often asked questions like, "So, which department is really responsible for the quality of the product?" To me, the answer to that question is as simple as these five bullet points:

- Marketing is responsible for knowing what the consumer wants and needs and, therefore, accountable if the quality of a product made to specification does not meet the consumer need
- R&D is responsible for translating marketing need into product design and, therefore, accountable if the product design cannot meet the consumer needs as specified by the marketing department

- Production is responsible for making the product and, therefore, accountable if a product is made that does not meet the specifications set by R&D
- QA and QC are responsible for monitoring and maintaining product quality and, therefore, accountable if product that is outside of the statistical limits for quality control reaches the consumer
- Warehouse or logistics departments are accountable if the product left the factory on specification but was damaged in storage or in transit.

The responsibilities as outlined above are crystal clear, and there is no wriggle room for passing the accountability. Of course we also need collaboration between departments to achieve success, and that is the subject of the following section.

The Need for Healthy Tension

Between every department there should be some tension. The reason for the tension is that each department has conflicting priorities and challenges, and each department should be pushing the other to be the best it can be.

Marketing wants the very best product, whereas R&D may not be able to find the materials to make the exact product desired. R&D will want to set very tight tolerances and specifications, but variability in the manufacturing systems might not be able to match those tight targets. Planning wants to operate with minimum inventories, which requires flexible manufacturing, whereas production has high targets for volume and cost that are more easily met with less flexibility. Therefore, there's a natural conflict between departments that creates stress and that challenges each group. If this conflict is handled in a negative way, it can lead to arguments and blame. If handled in a positive, collaborative way, this healthy tension can be effectively used to raise standards.

As we move on to the third of the questions for Step A: Aim from the Heart, we consider the ground rules to use to run the business. These ground rules, or principles, will guide you and, when effectively shared, also guide your team through even the most difficult situations.

How Could You Let That Happen?

"Change your opinions, keep to your principles; change your leaves, keep intact your roots."
Victor Hugo

Back to the family vacation, which we'll say is to the Caribbean. We want to ensure that we all get there together in a way that's safe and healthy, that the journey takes no more than 30 hours, and that we can all sleep for at least eight hours during that period of traveling. We may decide that the journey is part of the overall vacation experience, and that it should be immense fun, with experiences planned for every leg of the journey. We could insist that the all food available is enjoyable and nutritious as we travel, and that we will not tolerate any in-family fighting. Whatever decisions are made, it is essential that we all know the basis for the decision and, if anything goes wrong, why it happened.

Similarly, for your business, it's important to set the principles that will form the basis for your company culture, as well as the limitations of what is and is not acceptable. That will form the moral compass for your journey and ensure that, in progressing toward the destination, the principles of the leadership are maintained.

I believe most people are good at heart and want to do the right thing. When they make mistakes, it is often because the guiding principles they were following, or were supposed to be following, were not clearly established or well-enforced. This lack of clarity about the principles that guide the organization can lead to a lack of clarity in decision-making. We may all love to follow a general doctrine of "do the right thing," but in reality the more clear and clearly stated the company principles are, the more effective the decision-making will be at all levels of the organization.

The story below illustrates my own experience with being guided by very clear and compelling principles throughout my early career, which then provided a clear guide for my future businesses.

The Company's Principles

The main reason I joined P&G from university was their core principles and values. I still remembering opening the corporate brochure 28 years ago, seeing those values listed, and thinking, *That's the kind of company I want to work for.* Those corporate statements have become more widely shared, and I believe it has never been more important to do so. The guiding principles on which decisions are made say everything about a company. They also allow a company to self-check and self-regulate, as decisions that are out of alignment with the core values are easy to spot.

In this section, we'll discuss many possible guiding principles – specifically, those that I feel have had the biggest impact on the successful companies I've studied.

The Pallet and the Yellow Line

The most simple experiences can be the most compelling and memorable, especially when taught by a respected leader. What I'm going to tell you happened to me more than 25 years ago, and I can still remember it as clear as day. That experience has formed my benchmark for standards and provided me with a wonderful illustration of leadership.

I was in my first job after graduating. I was a fresh-faced 22-year-old and had been given a small manufacturing department to manage. The department had 18 employees and one line supervisor, all working on packaging fabric conditioner into cartons. The technology was new at the time and the business was in a strong growth phase. I was excited about my job, as our product was starting to be shipped around Europe and would be the only product of its type on supermarket shelves around the region. My boss, a man named Eric who was the division manager, was a highly experienced gentleman with a reputation for being a strict, no-nonsense type. He quickly took me under his wing. I was eager to please and eager to show that I had what it took to lead an even bigger division than the one I started with.

In preparation for a visit from our manufacturing director, I had helped the team clean the equipment and gotten the department looking organized and prepared. Eric, a thirty-year veteran of the company, had exacting standards, and I wanted his opinion. When I was sure we were ready, I invited Eric to do a pre-inspection.

In retrospect, I know that I was looking for his praise for our efforts, and I wanted him to be proud of the work we had done.

When Eric walked into the department, he stood there for about 30 seconds and then said, "Why is that pallet of material touching the yellow line? It should be behind the yellow line." He then turned and walked out.

I was stunned. I was angry. And for a moment I was deflated. Despite all the work we had done to be ready, Eric had focused on a tiny defect – and a meaningless one at that! The pallet being on the yellow line had no impact on quality, safety, or cost.

Immediately, I did what most managers do and passed the crap downhill, asking the machine operator to get the pallet straightened up and behind the yellow line. Then I stopped, suddenly struck by the realization that Eric had changed my perspective forever.

As I looked around the department, I saw defect after defect. Not only were pallets touching yellow lines, but labels on equipment were not squarely positioned. There was equipment that was rusty and dirty. Some spare parts were poorly organized. I saw tools that were damaged, liquids in the wrong containers, and documents that were out of date. Then I noticed product being manufactured that, although it met specifications, was far from perfect, far from what our customers deserved.

My mood changed from upset and deflated to a feeling of desperation. The tour from the manufacturing director was imminent, and I was the leader of a far-from-perfect department. My career was about to go down in flames. There was so much that needed improvement that the effort required would be endless. It would take everyone's support, and then maintaining it would require continual attention.

As I remember that moment, I can feel again the sense of panic that set in that day, and I can clearly recall the lesson my boss and mentor taught me: Continual improvement and total employee involvement are what life in manufacturing is all about. Then again, what else is there for a manufacturing leader to do every day besides that? And so we did what we could in the time we had before the big tour.

That tour with the manufacturing director went really well. He didn't see – or at least didn't draw attention to – all the defects that surrounded us. He thanked the team for their hard work and the improvements they had made. On the way out of the department, Eric gave me the tiniest pat on the shoulder and a knowing smile. Lesson learned!

The Transparent Culture

So now we know where we're going, we have a framework for the departments that will take us there, and we have set some standards and limits. The direction and expectations are clear. What happens next? We have the opportunity and obligation to set the basis for the culture.

Every organization has a culture, and it is rarely what the leaders want it to be. The simple reason for that is because the culture was never intentionally set. The culture emerged, often over generations and through several different phases of business. As a leader, you have a choice. You can accept the current culture – the norms and expectations within the business – or you can set the culture you want. Yes, it is as simple as that. However long it took to arrive at the current culture, it is possible to start a culture change immediately, as long as you are clear on the new culture that is required. This is, once again, the decision of the leader, and only the leader. It can be the leader of the entire organization or the leader of a specific group or department who sets the intentions for the culture.

As a leader of the organization, the culture of the organization emanates from you. What do you want? What do you expect? What are the guidelines for people to follow? How do you make things easy while ensuring the culture evolves in the way you expect? The answer is to be clear and transparent. In the next step of the ALIGN Process, we will explore communication, but first let's get clear on what we want to communicate.

In each successful organization, there are some critical elements that I have seen demonstrated time after time. These elements are covered in this chapter. Being clear on which of the following elements apply to your organization and how you want them to be implemented will make all the difference.

Area Ownership

If you want me to be responsible, you'd better tell me what I am responsible for. If you want to hold me accountable, recognize my achievement, train me to excel or take initiative to improve something, then I need to know the scope of my area of responsibility. I call this *area ownership*. If I want someone to take the actions of an owner, to treat their area as though they were personally invested in it, then I need to provide clarity about what their area is and the scope of their authority.

People don't need to be financially linked to a result in order to take ownership. Although financial reward can be an effective motivator, recognition is, in my experience, a greater source of inspiration. To build ownership there must be transparency. The individual needs to know their area of ownership, be supported to be successful, be allowed and encouraged to take action, and given responsibility for the associated risks and consequences. Throughout all this, they should be given clear feedback.

Every employee in a business should be given a unique area of ownership. This can be ownership for an area of the factory or a building, for a section of the equipment, or for a system or procedure. There should be clear boundaries. It should be clear to each person what they are getting, what they are expected to do with their area, and how what they have ownership for adds value to the organization.

Any decision-making regarding an area owned by someone must include the individual to who ownership has been given. This does not mean they make every decision, but they must be included in the decision-making. Just as you would not expect your own living space to be suddenly offered to others without consultation with you, the ownership of an area should be valued and treated with respect. Ownership builds accountability, which contributes to a desire for competence and leads to confidence. These are the building blocks toward mastery and should not be underestimated when designing an organization for high performance.

Personal Values

Every individual has a different set of values, formed over the years through their conscious and subconscious experiences. This personal set of values leads every individual to look at circumstances in a different way, through their own experiential lens. Discussing and discovering your team members' personal values can help you align the roles, responsibilities, and results of an individual with work that complements their values. In this manner, the engagement and probability of success in their role is elevated.

Aligning an individual's role and responsibilities directly with their values gives a much higher chance of both satisfaction and success. Imagine that high on an individual's values list was his children, but his role in the company required weekend work and doing evening shifts on overtime, so it was almost impossible for him to spend any time with his kids. Quickly, a conflict would arise, and he would be likely to get disillusioned with his work. On the other hand, for someone whose highest value was money, overtime work could be just the ticket to keeping her happy.

As leaders, if we help employees determine their own values and then flex their roles and areas of ownership to support those values, we will increase their chances of engagement.

A Learning Organization

Growth only comes through learning. Growth requires moving toward fears, stretching boundaries, making mistakes, and continual improvement. Business growth is no different. As we push the boundaries of the business we discover new issues and new challenges. We need an organization that is willing to accept and rise to the challenges, and is capable of doing so. Having an organization that is ready for challenges means leading in a way that is constantly about learning and about striving to get things right, even if not the first time.

To create room for new learning, the current learning must be ritualized into best practices and habits. This is the purpose of standard operating procedures (SOPs) and work rules. However, these best practices must be continually updated to reflect the new learning and new situations that naturally occur, so that they can be kept relevant. Out-dated SOPs are a noose around the creative neck of the team, so there must be a forum through which they can be constantly challenged and improved.

A learning organization is easy to recognize. Employees are encouraged to sign up for new training. Books, manuals, libraries, computer trainings, audiobooks, etc. are available. Employees are encouraged to take courses and develop in their areas of responsibility. After-work clubs are established and learning tips are displayed on notice boards. Knowledge trumps experience, and the phrase, "That's just how we do it here," is ridiculed. Audits are seen as positive events and outside guidance is considered beneficial. Interdepartmental discussions are encouraged, as are cross-training and the expansion of scope. Reward and recognition systems are tied to personal growth as well as to business results, and there is an energy and excitement pervasive in the organization. This is a great culture to clarify and implement.

Advance Warning – No Surprises

We all know that issues occur, but issues don't need to be surprises. I always see issues as opportunities to learn and grow. We face the issue head-on, find the root cause, fix it, identify the system that failed, fix the system, and then apply the fix to every other applicable case. This is how we make *issues* our friend. However, when we don't know there's an issue none of that can happen. When an issue is covered up, ignored, or not recognized until too late it becomes a surprise, and surprises are not welcome.

Most organizations suffer from a lack of effective communication. In order to rectify this, I encourage teams to over-communicate. This does not mean sharing every issue with everyone through mass emails with hundreds of CCs. Far from it. We need to communicate the issues with those who have responsibility for the issue and with those who will be negatively influenced by it.

As much as possible, this sharing of issues and their resolution should be through structured channels. A daily huddle or weekly review, if properly run, will catch most of these issues early, if the individual who has ownership for the issue area is fully in touch with their area of responsibility. When issues are uncovered, there is only ever one positive course of action. That is to foster transparency and work together to find the best solution for the business. In this scenario, there is no room for blame, cover-ups, backstabbing, or politicizing. Transparency is essential for finding the fastest and most effective solution. However, this does not mean there is no accountability. Once the issue is solved, there needs to be a review. In this review, the root cause and associated factors are discussed, including the systems that failed. At this point, accountability can be assigned and the appropriate measures taken to ensure that a similar event does not happen again (this is the purpose of any action or discipline).

Even when there is a surprise, the search for resolution is the same as above. But during the review there will most likely be more serious consequences – some for the issue occurring in the first place and others for allowing the escalation without asking for effective support. In a fully supportive culture, there is no place for surprises, and issue resolution should be carried out in a positive and relaxed atmosphere as a joint opportunity to learn and improve.

Gossip-Free

There is no situation in which gossip can be considered to be positive for the business or the individuals in the business. Gossip about colleagues, bosses, subordinates, customers, suppliers, etc. will gradually eat away at your culture. Yes, I know that gossip pervades society, that it is responsible for the majority of newspapers sold and articles written and read. I know that it is the mainstay of many people's entertainment and that it is often considered to be harmless and "just a bit of fun." But it is not.

Gossip is possibly the biggest cause of time-wasting, and it has the strongest negative effect on morale and is the biggest distraction from any business's mission. There is no such thing as positive gossip; there is always a victim, and whether they are inside or outside the company, the tone of the discussion is negative. Sure, the topic may be considered to provide "a good laugh" when it's about others, but what happens when it finally becomes directed your way?

Avoid the distraction. Build a safety net for your team by making the office a no-gossip zone. You will quickly find that the positive talk stimulates positive activity, that morale is boosted, and that trust is fostered. This is possibly the most overlooked element of culture, but it's one that is simple to stipulate.

Cleanliness

Another simple culture stipulation to talk about and to understand is cleanliness, but it seems that many people do not have the discipline to implement it. Cleanliness breeds clarity and must be implemented in all areas of the business.

A clean and clear desk is one where confidential information is protected, the user of the desk can concentrate on one thing at a time, and work is efficient. It amazes me when people allow cameras into their business, and we see a messy office and a desk piled with papers.

Such a scene screams of inefficiency. We see it often in the offices of doctors, lawyers, sales and marketing people, and in planning offices and production offices. What is worse is that some employees seem to wear clutter as a badge of pride. "Look at me," they seem to say. "Look how important I am with all of this work to manage. So much I don't have time to straighten up." Absolute nonsense. They are disorganized and inefficient, at best; most likely, they are not in control.

The same requirement for cleanliness applies to all workspaces, be it a maintenance shop, production floor, or warehouse. Clutter is a sign of a system out of control and a lack of discipline.

The need for cleanliness doesn't only apply to what we can see. Look inside desk drawers, tool boxes, and cabinets. Look on the tops of and behind machinery. Look in the furthest corners of the factory. There is never a good reason for uncleanliness or clutter. Elimination of gossip is a form of eliminating clutter from the mind. Eliminating uncleanliness and clutter removes inefficiencies and safety hazards.

Why do we find cleanliness and neatness so difficult? It's because our decision-making capability works like a muscle, and after a certain number of decisions in the day the brain becomes tired. Cleanliness is often left until the end of the day or the shift and, at that point, it's easier to throw all the tools in the same drawer, to leave the cleaning until tomorrow, or to stack the papers up rather than file them. This is a slippery slope. The only way to avoid those situations is to build time into the regular schedule and into the process systems to support area owners in maintaining clean and clear workplaces. In my experience, those behaviors need to be audited, until the owned areas are permanently maintained and the habits of cleanness and neatness have become part of the fabric and culture of the company.

Ground Rules

Why are manufacturing companies so often in a mess regarding basic ground rules and guidelines? It comes down to three major factors:

- Ground rules or guidelines were never set in the first place.
- The discipline required to maintain the standards until they become autonomous has not been provided.
- We live in a world of infinite distraction.

Set the ground rules. You are the leader and you have the responsibility to define the standards to which your business will run. Set the standards, write them down, and be transparent about why these are the standards you expect. Don't do this by committee. The committee will water down the standards based on the lowest standards of the members on the committee. This is *your* workplace, so set the standards for the company to follow. In terms of area ownership, you own it all. Set your standards, even if there is a gaping chasm between where the business currently is and what you expect of it. We will close that gap later. For now, you need to clarify and know your standards so you can assess reality against them.

We all know we can't change everything at once. Biting off more than you can chew is not a reason to lower your own standards in order to move forward. So don't do that. This setting of ground rules is where prioritization and discipline come into play. We will spell this out in more detail later, but for now, know that the ground rules you choose to set are about what to do, specifically what not to do, and what to say no to.

This is possibly the most important factor for leaders, and it's a critical element of the ALIGN Process. On a journey, you cannot make rapid progress *and* stop to smell every rose. You have to choose which roses to smell and which to look at longingly as you speed past. You have to have the discipline to keep driving, when that's required, and to stop when planned. Discipline involves changing your habits. It requires willpower until the habit works automatically. At that point, a new challenge can be undertaken.

In a world of infinite distraction, there is always something to do that will provide a more immediate shot of dopamine. We say, "Just let me check my notifications first," and then, "I've been called to an emergency meeting," and, "I'll get to that later, when I have time." When we're aware that there will be attractive distractions, we need to work out systems to keep us focused. One such system is called *forced optimisation*, in which a factor is put in place that is so compelling that the task cannot be avoided. Another system is auditing, in which a checklist for completion and compliance must be filled before something else can happen. You wouldn't want your pilot to take off unless he had completed his pre-flight checks. I wouldn't want my Area Three Leader going home before she had ensured that all tools in the toolbox were accounted for, clean, and in position, ready for the next shift.

Action Items for Step A

To guide you through this important step of building a foundation through the direction, responsibilities, and principles you choose to put in place, and in order to inspire and align your organization, the result of Step A must be absolutely clear. It will be the foundation for everything you communicate within the company, and for the vision and standards you will be remembered for.

To get started, answer the questions below from your heart. Protect yourself from too much "reasoning" at this stage. We can craft your foundational message for effective communication later. To inspire your organization, the message must be based on your heartfelt commitment.

Then it will be *your* direction and *your* wish for the company and for the future of all employees, that you share.

1) Answer the following questions:

- What is your compelling direction?
- Where are you going and why?
- What are the framework decisions you have made for your organization?
- How do you intend for the business to move forward?
- How do you think you will most likely get there?
- How will you structure the organization?
- What groups will you need at each stage of the manufacturing process?
- What will be the main roles and responsibilities of each group?
- Where will the boundaries between groups or departments be?
- What will be the operating strategy that guides your every decision?
- Who will you hire, and how will you hire them?
- What are the principles most important to you as you progress in this journey of aligning your business?

2) Create the Alignment framework:

This exercise takes you through a process to help you create the framework for your business's alignment. It will require some work to expose your inner drives. It takes a lot of courage to open enough to allow your inner direction to emerge. It is already in you, but it doesn't take thought to identify it. In fact, overthinking this exercise will make it less effective.

Do the exercise, then do it again, until you feel comfortable that you have reached a foundational starting point for continuing the ALIGN Process. Don't try to make it perfect. Don't try to force it. You will not get it right in one sitting. It will evolve, but go ahead and get started now. You will likely feel a range of emotions as you work through this exercise. If you get to the point of despair and giving up, then you are on the edge of a breakthrough. When you feel elated, then you are close enough.

Write down your answers to these questions:

- What is *your* destination and direction for your organization?
- What is the framework or foundation of your action plan – what are the areas of focus that will be required to drive the business forward in your direction?
- What are the ground rules and expectations your team will follow?

For help with this exercise, you can download a template for doing the alignment framework from www.makeitrightbook.com/bonus.

Step A Bonus Materials

Step A can be difficult because of two main issues: needing to detach from the current company norms in order to identify new ones, and knowing what is possible. Both of these potential challenges are explored below in the various ways I have seen them crop up.

These bonus materials give you insights into hundreds of other companies around the world as you get in touch with your unique mission.

Freedom vs. Security

Security is an illusion. I thought my job leading our family business successfully for four years was secure – until my father-in-law fired me! I never saw it coming. One day I walked into my office, and my desk was gone. Everything else was in place except my desk. Still not catching on, and with no one around me saying they knew where the desk had gone, I moved to another desk. The following day, that desk was gone. I started to get an inkling that something was wrong.

I was the managing director of the company, the eldest of the second generation to run the company, and a 20% shareholder. We were making strong progress as a business. Then suddenly I was out. The chairman, who was a majority shareholder, wanted me gone, and so I was. He didn't tell me his issues or his reasoning around the decision. There was no discussion that involved me. There was only a missing desk.

The most stable and secure companies are now crumbling. Countries are crumbling. Economies are shifting and becoming increasingly hard to predict. Holding on to some type of security is like trying to hold on to the mast of a sinking ship. You will still go down. Many people like a feeling of security, but the term *maximum security* takes on a different feel in these circumstances. Are you so "secure" that you are confined to a prison of your own making, restricted from going outside, limited in activities, and with a very specified agenda to follow?

An alternative to *security* is *freedom* – the freedom to set your own direction, to sail your own boat, to make your own choices, and to pursue your own dreams. Freedom comes with obvious challenges. The boat may capsize from time to time, and some dreams turn out to be nightmares, but the joy is in the journey. The wind takes your sails, the sun shines on your face, and you speed forward in a direction of your choice. This is what I live for. And this is what choosing your own compelling direction gives you.

Inspiration vs. Fear

As a young child grows, we use techniques to motivate them that increasingly tend toward control. Rather than encouraging and cheering every new step, we begin controlling the child. "Don't go there!" "Sit still!" "Be careful!" "I told you that would happen!" Gradually, we limit the risk-taking propensities and the creativity of the child, preferring that their actions are stable and predictable. Rebels are weeded out for punishment as examples to the rest. Our school systems are well designed for this, and they served the purpose, during the Industrial Revolution, of preparing individuals for factory work and doing prescribed tasks with minimum complaint.

As leaders of companies, we can become talented at using fear as a motivator and controller.

- "If you stay focused with us for 15 years, there's a good chance you'll become a partner."
- "Your bonus will be assessed based on your performance."
- "The economy is really bad right now, so we need even more from you."
- "The work rules are there to protect you."

Although I'm a big fan of using structure to promote efficiency, and of performance-related rewards, in my experience innovation and meaningful work stem from inspiration, not from fear.

Collaboration vs. Competition

There are two natural states of interaction between departments: competition and collaboration. Both can have positive benefits if leveraged successfully. Some level of competition can be used as a positive challenge, like when two training partners push each other to greater achievements. Negative competition or rivalry pushes both parties to underhanded methods and pulls the organization down. Gossip, backbiting, cheating, and misrepresenting are all signs of negative competition in the culture. There must be a zero tolerance policy around this type of negativity.

Collaboration is about talking to each other, taking on the challenge of a shared issue, and doing so in a manner that produces the best outcome for the consumer. This type of collaboration is not done by consensus, where each party is satisfied that the other is doing their best. Collaboration still requires each group to challenge each other and drive toward excellence, but when the discussion is over and the decision has been made, all must agree to support the decision. If you don't like it, talk some more; talk as much as is needed. Challenge, challenge, challenge – but walk out of that room aligned and moving forward.

Prevention vs. Cure

In many organizations, there is a bias toward troubleshooting. Those who troubleshoot issues are treated as experts and gurus and put on pedestals. This is not productive. Troubleshooting is needed, on occasion, to solve problems, but the main focus should be on *preventing* problems. Work on quality assurance, on ensuring that the conditions are right for making good product, rather than on quality checking to see if good product was made or not. Put the focus on doing preventive maintenance to prevent failures, rather than doing breakdown maintenance to fix machines after they've failed.

Although every organization has issues right now, there must be a strategic decision to break the cycle of troubleshooting and invest time and resources into preventing issues instead. This means working on areas like data analysis to predict issues; process control to optimize process settings; and cleaning to maintain equipment at the ideal condition. If you don't prioritize time for doing trouble prevention, you will spend more and more time on trouble fixing.

This sounds obvious, but it is human nature to fight against prevention. Think of your kitchen drawer, or that messy desk drawer. You know it will be more effective to keep it tidy, to have a place for everything and put everything in its place. So you arrange it once and decide to maintain it in that condition. Then, one day, when you are tired, you can't be bothered to put that thing back in the right place. Or you get something new that you don't want to "find a place for" right now, so it goes in the drawer. Very soon, the drawer system is back to square one.

Now imagine those good intentions in a high-pressure production facility with a variety of stressed employees working on shifts across the full day. With all employees responsible for a bit of the organization, do you see how quickly the "drawer" can get all messed up? Even with a perfect start and the best intentions, it's difficult to maintain orderly conditions.

This is why many of the production operations that I visit for the first time look like a scrap heap. It happens gradually, everyone gets used to it, and then it's out of control.

The way around this is to provide clear systems and to discipline with some kind of forced optimisation strategy. We'll discuss this more later but, for now, know that the way you set up your operation is the way it will work. To get the "drawer" cleaned up and to do it effectively requires the right operating strategy.

Optimized Equipment vs. Volume

Many manufacturing company leaders think their role is to maximize production volume. This may be the worst business operating strategy, and it has led to the decline of many businesses.

Why is maximising production volume an ineffective operating strategy? Let's say you're an operator on the night shift and your goal is to maximize production volume. What will you do when part of the equipment starts to shake? What will you do when part of the equipment starts to get a build-up of contamination? What will you do when the variability of one of your unit operations starts to get close to the limit? I'll tell you what you'll do, and I know because I've been there hundreds of times. You will continue running the equipment to maximize the volume, and you will hope that there is not a bigger failure as a result.

This desire to maximize production volume for the shift, day, or week is the biggest operating mistake I've seen manufacturers make.

Am I saying that wanting to maximize production is a bad goal? No! That should, indeed, be the goal, but maximizing production is a terrible *operating strategy*. Production volume is an output, not a strategy, and missing that concept is where so many businesses get it wrong.

A strategy is the way we will run the business on a minute-to-minute, hour-to-hour, and day-to-day basis. A strategy is based on how we will achieve the outputs in the most effective and sustainable way. No business is a short-term sprint. We need the strategy to keep us going through all circumstances – to protect us during the downs and support us during the ups.

In all of my work with manufacturers, there is only one operating strategy that works: maintaining the equipment in optimum condition to produce the optimum product.

What about the people who run the equipment? Well, their job is to have the skills and capability to maintain the equipment in optimum condition. What about the systems around the process? Exactly the same answer. *Everything* in the business should be about keeping the equipment in optimum condition.

This works because it's equipment condition that determines variability in product condition. Variability is our biggest challenge. Our goal is to reduce variability, and maintaining the equipment at the optimum settings is the way to manage this. Have I said that enough yet?

Equipment in optimum condition gives us:

- Lowest variability in product quality
- Lowest chance of process failures
- Lowest chance of equipment breakdowns
- Lowest cost of manufacturing
- Maximum uptime of the equipment
- Faster equipment changeovers
- Improved initiatives
- Less cleaning time

There are no negatives, which is why this is the right operating strategy. Why do so few companies use this as their operating strategy? It's because of human nature, which focuses on greed and short-term gains. It takes immense discipline to stop a machine that is running well in order to clean it. There is a huge desire to allow it to keep running *just a bit longer*, to squeeze *just a little more* from it, to push the boundary and get a few thousand more products out. This is the slippery slope to decline on which many manufacturers live.

The operating strategy must be directly in line with what we want in the long term for the business. The operating strategy must be so crystal clear that there is no decision to make. It must be engraved in stone and forcibly applied until it becomes the natural and normal way of being. This requires discipline, guts, and feedback. It requires being stronger than the excuses. And there will be many excuses that challenge this strategy.

- Just for this order!
- This is a unique situation!
- Our boss asked for…

Few of us are stronger than our excuses, but there are techniques we can employ to help ourselves stick with the strategies that we know are good for us. Forced optimisation is a strategy that works really well. Imagine a production line that is programmed to stop itself after every two hours of consistent running, so that it can be cleaned and maintained, and it won't be started again until a checklist of items has been completed. In cases where you can't program your equipment like this, then program the people. Put clear standards into place and then develop techniques to encourage employees to "do the right thing," and challenge them when they "do the wrong thing." This is where the feedback comes into play.

Helping a Few vs. Leading Many

There are numerous studies on how many directly reporting employees a leader can most effectively lead. There have been pushes toward flat organization structures, horizontal management across functions, and – more recently – organizations with no formal management. I propose that the initial question is wrong. Instead of "How many people can I effectively lead?" I prefer to address the question, "How many people can I directly and rapidly help when they ask for support?"

When posed this way, the answer directly relates to the capability and needs of the direct employees. If a team of experts in their field is employed, they may have little need for support, in which case a high number of those direct employees can be supported. When a team or individual has less experience or capability, a higher level of help and support are required. That necessarily reduces the number of employees that can be supported by a leader.

In all cases, the goal is to manage in a *transformational* not a *transactional* manner. As leaders, we should be helping and supporting, not telling or instructing. We should be working on the systemic and strategic items as much as the tactical ones. In my experiences with teams of all sizes and capabilities, I've noticed that, for me, there seems to be a sweet spot of between five and ten directly reporting employees. If there are more than ten, I find it hard to get to know the individuals in enough depth to really be understanding, as well as having not enough time to give the support that's needed. With less than five direct subordinates, I get a definite feeling that the organization could be flatter.

Helping the Best Excel vs. Helping the Worst Improve

It's easy to get into a trap of working with our worst people. You know the types – the ones who always have a problem for us to solve; who need time with us to complain about what everyone else is doing; who seem to cause more issues than they solve. Although I believe these people are only a result of the life circumstances they have been around, I simply do not have time to dedicate to them during the working week.

Is that a bit cruel? Shouldn't everyone be treated equally and given the time they need? No. It's my duty as the leader in the business to get results, to work with the people who will give us the best results around sustaining and growing the business. In this way, we help everyone, not only the individual. It's my duty to spend my time with the very best people, helping them utilize their strengths for the maximum benefit of the company. It's my duty to spend time to leverage the best people in order to pull the rest of the organization forward.

This issue is also role modeling the behavior I expect from my team members.

I have a zero tolerance policy for gossip, complaining, bitching, and moaning about what *should be* or how people *ought to* behave. People who have those behaviors and attitudes are energy drains. We need everyone's full energy and attention on growing the people and the business.

Individual Treatment vs. Equal Treatment

The subject of energy drainers, complainers, and negative influences brings me to a potentially divisive statement: I do not believe that all employees deserve to be treated equally. I do believe in equality. I do believe that all people should be treated as equals to me, but that doesn't mean treating all people the same. I believe in treating people as individuals, with their individual needs at the forefront. As I alluded to earlier in the book, everyone wakes up with a different head on his or her shoulders each day. No person is the same today as they were yesterday. No individual can be compared to another individual, because the ground is always changing.

The role of a leader is to tap into the individuality and diversity of a group to extract the maximum benefit for the business; to inspire the individual to use their unique capabilities to deliver excellent performance; to challenge, console, and encourage the aspects and behaviors that will move the individual forward to success.

Should I treat both of my sons the same? If one likes football and the other likes cooking, should I force them both to do both activities? Should I encourage either of them to turn away from his passion and aim toward what I think is "right" for him. Of course not. And the same applies to employees.

Our job as a leader is to find the true strengths in each individual and set up the environment for those strengths to emerge. And our job is to treat each individual in a way that draws out their strengths and allows them to shine. We cannot do this by treating people equally. That simply doesn't work.

Let's apply this concept to the energy vampires, those who suck the life out of us just by being around. We need to let them go so they can do their work in another environment. I don't ignore people – I always give them a chance to find and express their unique capabilities. But when it comes to getting stuff done, they either learn rapidly what I expect or they find an opportunity to go to a place where they fit better.

Give the majority of time to your very best people. Spend time building an environment where all can grow with their unique talents, but don't think that people "deserve" your time equally. Be clear about your expectations, about what you will and will not tolerate, and that will make this stance exceptionally clear to those around you.

Recruitment vs. Employment

People are everything in any business that relies on the frontline workforce to produce results. This applies to almost every manufacturing company. With people being the most valuable resource, it's important to pay extreme attention to recruits.

Some leaders pay so much lip service to the idea of people being the most important and valuable resource only because it sounds politically correct to say that people are more important than the equipment, the systems, and the buildings. But, from what I've seen, many say this, but don't believe it or get it enough to act on it.

I was once talking to a vice president of human resources for a major international company. I was talking about treating people as individuals, understanding their strengths, and recognizing and rewarding their contributions through regular discussions and appraisals. The more I talked about how we could tailor his organization to recognize the individual strengths of each person, the more I could see him becoming agitated and disconnected. So I asked him, "What is it? What are you feeling right now?"

"Kevin," he said, "you need to understand that I don't like people! I can't be bothered with all of this. I just want a system in place to get them paid and to prevent a strike."

Now that said it all. And this was a career human resources professional.

I do get where he was coming from, and, when I consult with manufacturing CEOs, I am sometimes called naïve. "Kevin, when you've been around as many types of people as I have, as many different characters, and for as long as I have, they'll wear you down, too."

Really? And whose fault is that?

I know that some people get energy from being around people (they are sometimes called extroverts, but I think that's an oversimplification), whereas, for others, being around people drains them. This is all the more reason for finding a niche that inspires and energizes you. Imagine if the person at the top of an organization that is driven by people says, "I just don't like people!" That person would be happier and more effective if they did something different.

What this has to do with recruiting is that recruiting should not be the job of the human resources (HR) department. I see HR's role, in organizations that are large enough to have an HR department, as providing a pool of candidates for positions. But it's the role of the specific leader of a team or group to recruit the right person for the position in his own department. My recommendation is to make it a significant aspect of the role of every leader to identify talented prospects and recruit them.

Recruiting should only be done by your very best people. One of my early managers, told me something that has proven to be correct time and time again. He said that "A" people (the highest performers) recruit "A" people, "B" people recruit "C" people, and everyone else recruits crap. There are several theories about why this happens, but the one I go with is that "A" people are not afraid. They are doing all they can to move the organization forward and they want the very best colleagues to work with. If someone else excels, they celebrate him or her because they see it as a benefit to the organization. An "A" leader wants the very best recruits on his team and will value the time it takes to recruit them.

"B" people are fearful that they will be replaced. They can't stand the thought of someone else being ahead and, therefore, pick inferior people to be around. That is self-defeating, as the whole division fails even if that individual "B" player still feels like the best of a bad bunch.

"C" people simply don't care enough to go through the process of assessing and finding quality recruits, and so they choose whomever is available. They don't recruit, they simply employ. Occasionally, they will pick a gem, but their process is more like a lottery.

Ensure that your very best people at each level are the only ones doing the recruiting, and ensure that they consider attitude as a major criterion.

Your Behavior vs. Quality Standards

I love the idea that whatever you are willing to walk past without taking action on becomes your standard. How does your behavior influence the standards? What level of quality should we be accepting, and why? The answers to those questions are critical factors in the success of a business. Of course, the quality specifications you choose to set are linked to market conditions, the needs of the consumer, and the role your company wants to play in the market. However, I make a clear distinction between *quality* and *specifications*.

Quality is how well you manufacture to the specifications. In this case, there is no tolerance for off-quality (out-of-specification) product. If you choose to release out-of-specification product you are saying, loud and clear, "The specifications don't really matter." The production department will translate that as, "We can make low-quality stuff, and it will probably get released."

"Hang on," you say. "We can't possibly only release perfect product. Didn't you say earlier that there was no such thing as perfection and that there's always variability?" Yes, I did. And that is what specifications are for. For each specification there is a target or goal, and there are tolerance limits. Although we always aim to run the process on target, we control our variability within the set limits. Each limit will have a dual consequence. The wider the limits, the more variability we will accept, the more product can be released, and the more likely the production department is to hit their volume targets in the short term; standards are more relaxed and, in the short term, the costs may be reduced, as there is less scrap product. However, the consequence for the consumer is that the wider variability means that they get more difference in the product on each occasion and that the function of the product has wider variability. Will the consumer notice this variability? That depends on the size of the limits.

This is a strategic decision on limits that has nothing to do with whether the product that has been made is good quality. If it is made within specification, as determined by the limits set by the leadership, then it is considered good quality product. The choice of targets and limits for the specifications will determine the position of the product in the market.

As we discussed before, the key to long-term success (equipment performance, high reliability, consistent quality) is in reducing variability. I have observed that companies with the highest quality standards, tightest limits, and a run-to-target mentality are the ones that eventually win in the marketplace. They spend their time and money on running with excellence and, in the mid-term, this strategy reduces all elements of cost.

Setting the quality targets and limits is the role of the leadership, with the support of the quality and process engineering groups. No one can make this choice except the leadership and, once those targets are set, there should be zero tolerance for off-quality product.

Your Action vs. Safety Standards

While we're on the subject of non-negotiable standards, let's look at safety standards. The conversation is almost identical to quality standards, and it is also a matter of life and death.

Each organization makes a choice about what level of safety is acceptable, and what the acceptable behaviors and tolerances around those behaviors are. It's all too easy to say, "We have a zero tolerance policy around safety," but that needs to be clearly specified and acted upon.

We need to make our way out of the realm of dreams and into the land of reality. Every action has potential consequences. We are humans and, unless we are completely inactive (which has the consequence of death), there will be opportunities for things to go wrong. The key in a manufacturing business is how we set the limits for the environment (workplace, materials, equipment, etc.) and behaviors, in order to reduce the possibility of an accident. Since there is the possibility of an accident (there always is), how do we limit the likelihood of an accident and the potential severity of an accident, if there is one? These standards around environment and behaviors are critically important when it comes to leadership and fostering a safe workplace. However, the actions taken when the standards are compromised are what determine the real standards.

Specifying acceptable limits for the work environment and behaviors is a major role of the highest level of leadership, and it cannot be delegated. Managing the actions taken to keep the company within those limits, with a zero tolerance policy for environment situations and worker behaviors that are beyond the limit, is critical. Once again, what you are willing to walk past without taking action becomes your standard and your expectation. Actions always trump the set standards.

Always New vs. Shiny New

I love to see new equipment come in to a factory. It is generally a sign that the business is growing and that the leadership team are investing. Whether they've invested a few thousand or a few million dollars, there's an immediate shift in behaviors. The equipment is carefully unwrapped and positioned in place. It's properly aligned and carefully started up. Photos are taken, as the team is proud of their new asset, and then the equipment slowly gets started up. If the organization is relatively advanced, there will be a commissioning and qualification procedure wherein the teams are carefully taught how to most effectively start up and run the equipment. There may then be a validation process, to ensure that the teams really can run the equipment at its ideal settings to produce the expected results. This is a wonderful period of group collaboration, aligned objectives, and controlled project mentality.

You can think of it as similar to buying a new car. With pride, you take photos, ensure that there is a sufficient and nice place in the garage where the new car will be parked without damage. You read the manual and commit to having the car serviced. Each time it rains, you wash and polish the car, and perhaps you even leave the plastic on the rear seats for a while. Before climbing in, you shake the mud off your boots and make sure there is a suitable place to keep your coffee without it spilling.

And then time passes. Hedonic adaptation sets in. The new car becomes your new normal. Your brain adjusts to accept the new reality, and it's not quite as exciting. Gradually you start to leave your car outside rather than taking the time to put it in the garage. A few more weeks go between polishes than when the car was newer, and the mats in the front are getting used to being dirty. The excitement and sheen is wearing off both the purchase and the car.

Now transfer this behavior to a factory. That new equipment was never given an owner and the standards for its care were not clearly set. Four different teams ran it across three shifts, 24 hours per day, 6 days a week. Once production pressure came on, there was less designated time for cleaning, and the new size-change part that was bought didn't quite fit properly. In fact, when the size change was done, some of the washers went missing from the nuts and weren't replaced. In the rush to get the equipment running again, the new size-change part was nudged into place using a hammer – just a few quick taps to get it to fit. A notice was then stuck onto the guard of the equipment with tape to indicate caution, as the guard would no longer shut with the modified size-change part.

The machine is no longer looking or operating like new. Imagine your new car with tape stuck to the doors to cover scratch marks, hammer dents in the wheel rims, and a build-up of grease and dirt on the engine block that gradually gets burnt on. The beautiful new Ferrari now starts to resemble an old banger.

This damage is rarely seen on a day-to-day basis, but it always happens day to day. The small, wrong decisions based on a lack of attention to detail, lack of effective problem-solving, lack of support, and lack of enforced standards creep into the daily workings of the plant. As standards slip and are not noticed and addressed through effective systems, control is lost. It's not long before operators are whacking the new Ferrari with hammers and chisels, and the cost of that damage is charged to the consumer.

The solution is to rigorously maintain the equipment in the ideal condition.

Bolts vs. Tape

I use the term "tape technology" for any "fix" on the equipment that does not restore it to initial conditions. Like the creep in standards for quality and safety, this tape technology is a gradual cause of increased variability. Sticky tape is used as a quick fix. Imagine that a small bolt in the equipment falls out and goes missing. The operator goes to the storeroom to get a replacement bolt, but there isn't one of the right size available. In order to run the equipment on his shift, he decides to wrap some tape around the bolt as a temporary fix, thinking that the day shift can replace the bolt. On handover to the next shift, there were some other issues, so he forgot to mention the sticky tape and the bolt. Over the next three days, the tape remains, but the bolt gradually starts working loose from the vibration of the equipment. Rather than fix the root cause of the problem, more tape is added. During the next size change for the equipment, the gap of the equipment cannot be effectively adjusted, because that bolt is missing. The operator decides to fold a piece of cardboard and strap it onto the equipment with some additional tape.

Does that sound ridiculous? What's most ridiculous is that I could walk 99% of production facilities around the world and see that exact situation on the equipment. Don't let it happen on yours. Tape technology is never an acceptable solution and is a standard that must be realigned.

Restaurant vs. Canteen

Have you heard the saying, "An army marches on its stomach"? Based on the canteen foods in many factories, you would think the company wants the army to roll over and go to sleep. And that is the daytime food that's available when the office staff and dignitaries are around. On afternoon and night shifts the food is often a disgrace.

The frontline workers are your eyes and ears on everything that happens at the plant. They control quality, cost, safety, inventory levels, and production output. You want inspired, active, capable individuals who are taking initiative and driving the business forward, don't you? You want people who are thinking *quality, quality, quality* in all they do. A great opportunity to reinforce this message is in the quality of the food you serve them – and the restrooms you provide. Every employee will need food and restroom services at least once per day.

I've heard all sorts of dismissal and evasion when I've brought this up with clients, ranging from the obvious – that it's a cost issue – to the downright ignorant – "These are just factory workers. They feel more comfortable with it like this."

To keep your team alert, they need great-quality, healthy food. Having a meal is often the one time in the day they get away from the noise of machinery, the pressure of production, and the demands of their teammates. Allow them time to recuperate in a calm environment with nourishing food. Recognize that they are your greatest assets, and so they deserve to be treated as such.

Visit the restrooms to see if *you* would like to use them. Often, all that's needed is a lick of paint, maybe a few new tiles, and a cleaning maintenance routine that keeps them fresh and dry. Get the best toilet paper, hand soap, and hand dryers. Treat your workers like you want to be treated. (If you are a customer of a company and doing a factory tour, ask to use the facilities. They're a great reflection of the quality of product you will be purchasing!)

Today vs. Show Day

I can count on one hand the companies I know of that don't need to do special "visit preparations" before a senior manager arrives. What a waste of time it is to have to use dedicated resources to clean things up for a visit from a senior manager. What an insult it is to the people who use the factory every day that the daily standards to which they are subjected are "not good enough for the boss."

Here's a different concept. Imagine that everyone was so proud of the condition of their factory, every day, that they encouraged visitors, without needing any preparation for them. Imagine if the Prime Minister or the President showed up at the door of your factory and requested an immediate tour. Let's go a bit further. Imagine that you wanted to take your parents or you children around and show them where you spend a significant percentage of your life. Would you be able to do that in your factory today, right now, with a sense of pride?

Being visit-ready every day is not only about cleanliness. It's about showing efficiency in everything you do, all of the time. It demonstrates effective systems, equal respect for all people, and a high regard for safety and quality.

Life Balance vs. Work-Life Balance

I used to be quite a proponent of work-life balance, on getting home from work on time to be able to live some of my life. Leaving work after having spent eight, ten or more hours there, to spend only a few hours on life before I went to sleep, made me realize how insulting to myself the term *work-life balance* was. It suggested that while I was working I wasn't living; while I was working, my life was on hold, as though I was generating income at work so that I could enjoy the life part of my existence.

I did the math. When I added up getting ready for work, traveling to work, doing work, and going home from work, it balanced out at about 12 hours per working day, which, for me, was 240 days per year. That meant I was spending 2,880 hours per year working – 33% of my total time.

With seven hours sleep every night I slept 2,555 hours per year. I spent 5,435 hours per year working and sleeping, which meant 62% of my time was gone. That left 38% of my time to "live my life."

I don't want only 38% of my time for living my life. *Work-life balance* sounds like a seesaw, with work on one side and life on the other, in competition. We are left trying to seek a difficult balance, one where work doesn't eat into our life or where we don't work too little to be able to have the money to enjoy life.

I want work to be part of my life, an enjoyable, fulfilling, worthwhile element of my life. I want it to be an area of my life where I am challenged and I grow, where I strive for greatness and am recognized for my achievements. I want work to be such a fulfilling part of my life that I have to improve the other areas of my life to match the enjoyment of work.

As leaders, we have the opportunity and the obligation to eliminate the need for people to strive for a work-life balance. I believe that a core part of our role is to bring this level of growth and joy to the daily work lives of every employee, to inspire them to reach greatness in their roles, and to expand their capabilities to the level of mastery.

Rather than setting an objective in the organization to have "a good work-life balance," how about setting the objective to allow every person in the organization to love their work, to see their work as a fulfilling and worthwhile part of their life, and to strive for greatness.

Summary of Step A

Step A on this journey is to develop a clear and compelling direction. That becomes your "why," your main talking point at every event, the reason for every review, and the yardstick by which you assess the alignment of your followers. It becomes your benchmark as you interview people and as you assess action plans and the alignment of your divisions. This compelling direction will drive every business decision. Yes, it is that important, and so you need to get it right.

Additionally in Step A, we went through a sample framework for making critical decisions about roles and responsibilities, boundaries and specifications. While the framework is, and must be, unique for each organization, there is a simple factor that is the same, and that is to clarify and set clear roles, responsibilities, standards, and limits so that the organization, teams, and workers know what is expected. Then they will be able to be held accountable and recognized effectively. One role of the leader is to set those standards, and that cannot be delegated.

Step A: Aim from the Heart, forms the basis for your direction, for the roles, responsibilities, and standards to which you will all work.

Next, we'll ensure that everyone knows about them.

CHAPTER 4

L: Lead with the Frontline

Step L in the ALIGN Process is Lead with the Frontline. It ensures we engage the critical decision-makers in building the detailed action plan, and that the priorities are aligned at every level of the organization.

This chapter explains how and why it is critical that you:

- Communicate the destination, direction, principles, and framework of the plan in a manner that inspires and focuses the organization.
- Enroll the frontline employees in developing the detailed action plan.
- Set the priorities in a manner that is transparent and compelling to all.

In this step, we discuss why getting the input of experts (frontline employees) is critical in building their ownership and commitment to the plan. We want to ensure that these experts are continually building their own knowledge base and staying ahead of the learning curve, because that is essential to finding the most effective solutions. Unleashing the energy of individuals, while narrowly focusing it, will provide the best solutions and the fastest progress toward the destination.

This chapter also explains how to pull together the master plan; how to get the most effective alignment and support between groups with sometimes opposing inputs and even conflicts of interest; how to ensure that every group or department understands the unique role and responsibilities of the others so that, once the plan is aligned, there is optimum help and support, and competition for resources is replaced with collaboration in the desire to deliver the best overall result. Finally, we will look at how to set the right key performance indicators (KPIs) for each group, such that we can measure progress and ensure that, throughout the journey, the groups work in synchronization.

This step is separated into three main questions that crop up for manufacturing CEOs:

- "What do you mean you didn't know?"
- "What were you thinking?"
- "Who agreed to that?"

Let's start with the first question and explore how to ensure that everyone knows what you plan to achieve.

What Do You Mean You Didn't Know?"

"Leadership is not about a title or a designation. It's about impact, influence and inspiration. Impact involves getting results, influence is about spreading the passion you have for your work, and you have to inspire team- mates and customers."
Robin S. Sharma

It's difficult to take a group on vacation if no one knows about the trip. It's difficult to get help and support for all the complex planning elements if the plan is not shared. It's hard to enlist full support if the destination and direction are not shared in a way that excites, challenges, and intrigues the supporters.

Addressing this first aspect of Step L is essential for coordinating all the resources that will make your mission a reality, so that you all know exactly what you will achieve together. It creates a sense of involvement and ownership when people know the specific roles they will play. This section is all about communication and *your* responsibility to align the team.

"I love it when a plan comes together" was the catchphrase of John "Hannibal" Smith, the Colonel played by George Peppard in the '80s TV series *The A-Team*. I know that feeling and, hopefully, you are starting to feel the same way. By following Step A of the ALIGN Process and answering the exercise questions, you now have a clearer destination and direction for the company; you have the guiding principles and framework of how to get there. And so it's time to enroll support. The supporters are not only your organization but also your suppliers, customers, shareholders, and the community. All will need to be on board with your plans and will need to understand with crystal clarity what you expect from them.

At this point, you should expect to start to sound like a broken record – although a very inspiring broken record – as you will now be moving into repetition mode. Repetition is a characteristic of how great leaders communicate, and your clarified destination will form the basis of almost every conversation. We will talk about how to most effectively communicate, and how to make those conversations inspiring. The most important thing is for you to feel deeply attached to the message you are sharing and be authentic in your delivery. After all, these are *your* direction, *your* plan, and *your* principles. If, at this point, you don't feel committed to your message, then don't share it! Go back and review Step A. Look again at where you are taking the group. The only way to inspire others is to feel genuinely inspired yourself. Once you've got that, it's time to share.

The Town Hall Farce

Times are changing and methods of communication are changing even faster – just ask a teenager how often she uses her phone to actually talk rather than to communicate in a multitude of other ways. It's time for CEOs to adapt and use the latest technology for spreading their messages. The story below indicates some of the challenges and the incredible benefits of moving with the times.

It's early in the new year and I know what's coming – the annual address from the CEO to the troops. This is one of the most painful activities of the year and of every quarter. Of course the Annual Address is only once a year, but then there are quarterly updates, those magic moments when the CEO reveals whether the company is doing well or not. And that sentence there says it all. Waiting for an entire quarter to know if you are succeeding seems ridiculous. But there's more. We run a manufacturing business that operates 24/7, 365 days per year. For the CEO to come talk with four teams working on shifts, this must be a major operation. As the discussions begin, will we shut down the lines for the talk, or talk to the teams on overtime at the beginning or end of their shift? What about the team that is on a day off? Do we bring them in from home? Shift handovers are at 7 a.m., 2 p.m., and 10 p.m. Will the CEO be able to attend at those times? Will he want to? Or should we split the announcements over several weeks, so he can do them all at 2 p.m.? For those who have to work on overtime or come in during their days off, will the CEO's address be mandatory or optional? What is the legal implication of making it mandatory? What if they don't want to hear the address? How then will they get updated on the plans and priorities for the upcoming year? And so it goes, on and on. The only certainty around this whole event is that it is incredibly time consuming and a distraction to running a production operation.

So should we cancel the CEO's address? *Absolutely not.* It is critical that every employee hears directly from the mouth of the CEO. What's even better, is that they *see* the words coming from the mouth of the CEO. And that is where the magic starts!

Back in 1920, the only way Henry Ford could talk to his operators was to shout at them through a megaphone. Knowing it was important to communicate to them, I feel sure this's what he had to do – stand on a stage and shout.

In the last century, technology has moved on, and I'm not talking only about the microphone. The issue we have is that the CEOs, HR department staff, and others involved in the annual and quarterly address debacle don't seem to have moved on at the same pace as the modern world. If most CEOs were 12 years old, they might get it.

We now have access to something called *video*. What's more, it's available in a format that can be beamed directly to mobile phones, the one gadget you can be sure that 95% of employees in almost every factory worldwide possess.

In a video, people can watch the words coming directly from the mouth of the CEO *at a time that is convenient for them*, at a time when they're awake and alert, not nodding off after a hard-worked 10-hour shift. They can access the talk in their first language, through subtitles and translators. They can really see the graphs and charts presented, and pause if they want to study the numbers more closely. What's more, they can have an updated message sent to them from the CEO every day, if it's available and if they want it.

Communication

In this day and age there is no excuse for an employee to feel out of touch with the organization. The informed CEO can send a one- to two-minute update every day to the entire organization and it needn't take more than five minutes of your time, if you're the CEO. Imagine the power in that, in directly communicating to every employee, every day, about the direction of the business, the culture, the ground rules, the progress, and celebrations of achievements. Imagine that, as a frontline worker, your team is mentioned by the CEO for the outstanding achievement of yesterday's efficiency result, which helped alleviate a situation for the sales team. Imagine the feeling as the CEO mentions your name on the address as a shining example in safety, or because today is your birthday.

The right communication fosters relationships, creates culture, reinforces the best behaviors, and inspires the team.

Constant Repetition

Over the years, there have been some masters of communication. Winston Churchill comes to mind, for his addresses to the nation during the war years. John F. Kennedy and Martin Luther King Jr. were also masters at communication, and they were limited to radio and TV for spreading their messages to a wider audience. What could you achieve with a direct and immediate line into every employee's phone?

Each of those masters of communication I mentioned had a few traits in common. One was the use of repetition. A phrase heard often enough starts to become reality. In all ways, you want your vision for the company to become reality. Think of the major brands and what they stand for. We associate Volvo with safety, for example, because that connection has been constantly repeated.

It is critical that, in your company, you constantly repeat the direction, the steps along the way, the importance of collaboration, and the specific needs of each department.

What Were You Thinking?

"The essence of competitiveness is liberated when we make people believe that what they think and do is important – and then get out of their way while they do it."
Jack Welch

I've never really been a fan of travel planning. I generally know where I want to go and why. I know how I want to travel and what makes me feel good. Heaven forbid I work through all the options of flight scheduling, whether to use airline points or cash, the ideal car to book at the best price, or whether the hotel has the right pillows so I can sleep soundly. That level of detail is so far out of my interest level and skill set that, if I ever need to do it, I get caught in loops of procrastination. It's not an exaggeration to say I hate that type of work and am willing to pay others to take the chore off my hands. What is also true is that planning an effective trip with all the elements expertly handcrafted to deliver a unique experience is a wonderful joy and passion for some people. *Those* are the people I want planning the details of my trips.

Your role as a CEO is to set a clear and compelling direction, set the structure of the plan, and set the ground rules the organization will follow – then *attract the right experts and allow them to work in their areas of expertise to deliver the company to the destination.* Your role at this stage is to get input from the experts about how they believe you can most effectively reach the destination and how they will work with others. It is also to listen to what help and support they need from you. Your role is to seek a wide range of input that fits the criteria and to encourage the experts to think outside the box. It is to challenge some of the "norms" that have gotten the company where it is. Your role is also to push the experts to stretch their own expectations, to stay focused on their areas of responsibility, to demonstrate what could be achieved with specific action steps, and to set priorities based on what will have the greatest impact for taking the organization in the desired direction.

Daily Decisions

There is a common fallacy in business that the managers decide what action will be taken and make the major decisions that decide the fate of the company. This is reinforced when managers get huge payouts if the company does well. I'm here to tell you that managers do not make the most important decisions. *The critical decisions are made every second of the day by the frontline employees.* This may be in the production department, where the frontline employees decide whether to keep the line running or stop it, whether to acknowledge the quality issue or let it pass, and whether to clean the equipment or let it deteriorate. It may be in the sales department, in the tone of voice used by the sales rep, the quality of the documentation completed, or the price offered to a favorite customer. These are the decisions that make or break businesses, and they are made by the frontline employees. These employees are more correctly described as *frontline decision-makers.*

Because these are the people most in touch with the critical elements of the business –production and sales – they should be setting the action plan. With direction given by the leadership team, these frontline employees will know best how to achieve the targets that will match the direction. And so we need to ask them what they think should be done!

Action planning is often regarded as a task only for the leadership. I disagree. The people closest to the decisions should do action planning. They know the equipment and the customers most intimately.

This section is not about *doing* the action planning yourself; it is about *assigning* the action planning to the frontline.

I am a realist. I know that not all frontline employees have been trained to put together a comprehensive and effective action plan. I also know that having thousands of people contributing to an action plan can be a challenge. Therefore, I suggest working with a small cross-section of the employees who have not been delegated by the masses but chosen by leadership as being the most capable – the ones considered to be the top five frontline employees of each group.

They will need to be guided through this process by a leader who knows how to listen, enroll, support, and ensure that the action plan will most effectively take the business in the desired direction.

Using the action plan framework that you developed in Step A, we start with the production and sales teams, the two core sections of the business. These two departments produce the product and sell it, bringing money into the business. In most manufacturing companies, this is the only way money comes into the business.

The production and sales departments construct their draft action plans and share them with leadership to ensure understanding and alignment to the business direction.

Once alignment is achieved, the plans are shared with all the support departments – engineering, R&D, quality, marketing, finance, etc. –

and those departments put together support plans that will allow the production and sales departments to succeed.

This entire process can be completed in about three days and is guided by Step L.

This is not the final plan – there is still a need for prioritization, alignment, and budget-setting – but from this process comes an overall sense of what the employees believe should be done to maximize opportunities for business growth and stability.

Alignment and Commitment

Why would we go through the headaches of forming a frontline action plan when it's clearly easier to have the leadership form the plan and deploy it? There are two main reasons: alignment and commitment.

1. **Alignment** – In every struggling business I have seen, there is a disconnect between the frontline employees and the management. That gap means management doesn't know, in detail, what is happening on the production floor, in the minds and hearts of the frontline workers, and, therefore, they are unable to set a plan that fully engages those critical employees.

2. **Commitment** – There is more commitment for a plan that is developed than there is for a plan that is deployed. In the process of developing the plan, the representative frontline employees build their own commitment to the success of "their" plan. This level of ownership of the result; this understanding of how and why the result is important and how it will benefit the entire company, is priceless. These employees will drive from the frontline, on weekends, and nightshifts. They will encourage and rally their teams like no voice from the office can do.

Allow the people who know best to set the plan. Allow them to recommend the priorities.

The Laser Analogy

There are two ways to manage large groups of people. I call these ways *the dimmer switch* and *the laser*.

1. **The Dimmer Switch** – This is the most common form of large group management. When the activity level in the group goes up, fear sweeps through the management. "Will this high level of activity result in a strike, a demand for higher pay, a loss of control on the production lines?" To keep those fears in check, management uses the dimmer switch. They find ways to reduce the energy of the group, to subdue the thought processes, to "keep in control." Performance-related pay, annual bonuses, lack of information, lack of leader involvement, and reduced ownership are all ways of controlling a workforce, of keeping them in line and keeping them dumbed down. (Some leaders also employ crappy food and unsocial shift patterns.) Reducing the energy of the team allows management to stay on top of the pile and to feel like they're keeping control. This is a widely employed strategy.

2. **The Laser** – With a laser, the higher the energy, the more effective the result – but only when it is accurately focused and targeted. Using the laser analogy, we stimulate the energy in the team. We encourage ownership, decision-making, knowledge, activity, prudent risk-taking, conversation, challenges, and dissatisfaction with the status quo. And we keep pumping in more energy. The key is to direct this energy with laser focus toward a specified direction for the

business. Massive quantities of energy focused on the most decisive tasks will take the business forward. The energy of the frontline teams becomes much more effective as it is guided by the company's clearly defined direction and supported and boosted by the support departments and the management. This powerful beam of human potential is what allows us to cut through challenges, to flow forward at the speed of light. As we celebrate progress, the energy increases. This is the key to exponential growth.

Encourage energy, and use the tools in the ALIGN Process to direct this energy accurately and acutely toward business success.

Creating Ownership

In the previous section titled "From Breakdown to Breakthrough" I explained that there was a time early in my career when I was in a learning phase and wanted to know all I could about how the production line worked. I asked hundreds of questions a day, studied, and learned how to do the roles of all the team members who worked for me. Within a short period of time, I thought I was the bee's knees. Each day, I spent 12 hours on the line. I loved it. I saw each of the three shifts that worked that day, and I was out there operating with them. I'd sweep the floors, load the materials, lead the troubleshooting, and support the size changes. There wasn't anything I couldn't do, and I was involved in everything. I was having fun and learning so much.

But there was a major issue with that situation: The line wasn't running well! Despite all my efforts at overseeing all three shifts, guiding the team leaders, and getting involved, the results were poor. I didn't understand it. Okay, there were some silly human errors made by the team members. Some problem resolution took longer than I'd expected.

And the morale of the teams was low. But I didn't think any of that would make the line run so poorly.

A few months later, I was seriously tired and was no longer enjoying myself. I was working harder than ever, for longer hours, and the line was *still* underperforming. We were starting to suffer from a bad reputation. What on earth was going on?

I realized, after suffering a personal breakdown, that the poor results were due to *my* actions. In doing everyone else's job, I was *stealing their meaning*. I was stealing their ownership and accountability, their sense of pride and success. I was stealing their opportunities to make the decisions, to take the risks, to feel alive and worthwhile. And so I vowed to never let that happen again. My role was not to be the best machine operator, the best maintenance technician, or the best process engineer. My role was to ensure that *the team* became the best; to ensure that team members were successful; to allow them the opportunity to learn and grow; and to direct their efforts toward making a difference in the world.

That single realization set me free. I immediately started working eight hours per day. I asked the team leaders what they needed from me. They set the action plan with their team members. They assigned their resources and set their own goals and targets.

It takes guts to allow the team to set their own action plan. It takes guts to step back and offer direction and support rather than answers. It takes a leap of faith rather than turning a blind eye. Don't steal the ownership and accountability from your team members. Allow them to succeed, and support and cheer them on their journey. Allow *them* to be the heroes.

It's All About *Us*

This business is not about you; it's all about the other people in the company, and you are at the heart of that. When you make the others successful, you have the maximum opportunity of being successful. You may be able to bluff a situation for a year, but after two or three years the organization will mirror your behaviors. If you make the business about yourself, it will be limited to your time and capabilities. If you are finding yourself in meetings that you don't want to be in, making decisions that you feel someone else should be making, and working long hours because no one else seems to be able to get it right, then you are in a trap of your own making.

But if you make the business about everyone else, then the potential is the combined potential of the entire group. It may seem that there is more risk associated with this approach, but the real risk is in limitation. The safe, self-limited companies are rapidly usurped by the unlimited companies. Who would have predicted the almost unlimited rise of Google, Yahoo, Facebook, Alibaba, Uber, Airbnb, or Tesla? It happened because the potential of the people in those companies was not restricted by the leadership, but was encouraged to flow and was directed toward clear and passionate causes. The leaders were not self-important (there is remarkably little ego in the founders of those companies). The missions of those companies, and the successes of the people, were everything.

Who Agreed to That?

"The role of leadership is to transform the complex situation into small pieces and prioritize them."
Carlos Ghosn

We have suggested plans for each step of our journey to the Caribbean. We have detailed taxi, bus, plane, and car hire options for each traveler. We have suggested accommodation plans, along with all the pros and cons of each. We have structured itineraries based on the needs and desires of the individuals, and their associated costs. Each group that is planning this trip has given their very best input. Now we need to make some executive decisions. How do we pull this all together? How do we align on the ideal plan that will best meet the needs of the individuals, with their varying desires, budgets, and constraints, as well as move us in the compelling direction that has been set by the leader? How do we find the plan that will give the maximum value experience for the entire group?

Every manufacturing organization is complex. Each department will have conflicting needs and priorities. Take, for example, the complexity presented by having multiple product lines and variants, often described as stock keeping units (SKUs). Having more SKUs is desired by the sales department, to give the customers maximum options, but multiple SKUs are a challenge for the production group, which has to make regular product changes. The planning team has the objective of meeting the needs of the sales department, with minimum inventory, but the production team wants longer production runs in order to reduce the manufacturing costs. Conflicting priorities and tension are inevitable, but they don't need to be a problem. It is the role of the leadership to make priority calls, based on all the input, and to ensure a balance that is most effective for taking the business forward. Not everyone will be happy at the end of this process, but they must all reach alignment.

The KPI Dilemma

Every year I see and hear the same story. "We have to set the KPI for our teams, and we don't know what they should be!" Let's look at an example.

Think of the key performance indicators (KPIs) for Usain Bolt, the Olympic multi-gold-medal-winning sprinter. For Usain to have ownership and commitment to his training schedule he needs to have set his own goals. He will decide if he is going for gold in the next Olympics. He will set targets for his personal best performances throughout the year. He will set how many early morning training sessions he will do and, with the guidance of his team, how much weight he will be lifting. The power of setting his own goals is that he has already completed the mental work around knowing what he is willing to give. He can thus activate his own inspiration.

KPIs are not for the business, they are for the people. They are set by the people, for the people, in order to provide personal challenges.

Since the individuals set their action plan, they also need to be setting their KPI. This is their own personal measuring stick – not that of the company.

Of course, the company will also have measures for achievement, but they will be compiled as a result of the challenges the teams have set for themselves.

Are you worried that individuals and teams will set their own goals too low so that they can be easily achieved? Step L in the ALIGN Process takes us through challenging, aligning, and prioritizing the action plans of each department and consolidating them into a master plan that the whole organization can align with. This is crucial for helping everyone be clear on who agreed to every decision, and why.

The One Big Goal

There can only be one goal for the organization, and this goal is the basis for every decision made. This goal provides the clarity required in every situation to make the "right" decision. There is only one person who sets this one big goal, and that is the CEO. Without this clear big goal, everything is debatable. *With* the clear goal, there is simplicity, a guiding light, and a determined outcome.

Imagine that the production team has made $300,000 of off-quality product. It's not all bad quality, but perhaps 20% of the total production lot does not meet the specifications. What's to be done about this?

- Do we rework the entire product to sort out the 80% that's good, repackage it, and sell it, and discard the 20% we found to be off-spec?
- Do we scrap the entire lot, thus throwing away 80% of good product?
- Do we sell the product and then deal with the complaints we receive if and when consumers complain?
- Do we find a lower-quality market to sell the product to at a lower price to try to reclaim the cost of materials?
- Do we implement a sampling program to try to increase confidence that some of the product has fewer defects?
- Do we do a combination of the above?
- Does it all come down to an ROI (return on investment) calculation?

There is only one answer to this: It depends!

It is at times like this, when there is a big decision to be made, that the one big goal comes in useful. Although it will not directly answer the question, it will cut through a lot of interdepartmental discussion. There is no doubt that the production, quality, sales, and marketing departments will have different perspectives. There is a perfect opportunity for blame, finger-pointing, and confrontation. None of that changes the fact that the situation has occurred, and it needs to be dealt with.

On every occasion when there is the natural opportunity for conflict between departments, there needs to be a guiding light that is directly aligned with the direction and goal for the business. The more specific and targeted the one big goal is, the more effective it will be in guiding the decision.

Healthy Tension

I mentioned healthy tension earlier, but it's worth revisiting. Healthy tension between departments is just that: healthy. Each group pushes the other forward and challenges the other to achieve more, deliver more, and stretch its boundaries. There are naturally different silos in organizations. Each group has its own budget, each has a loyalty to its own people, and each has goals it is striving to achieve. The role of the leadership is to align those departments with the one big goal that will drive the business forward in the most efficient and effective manner. This section of Step L is all about aligning the action plans of each department to achieve that goal and reduce the silo effect.

Win-Win Solutions

The one big goal is the rallying cry that allows competition to become collaboration. We want to eliminate the win-lose factor of competition but retain the element that spurs us all on. How do we fight to be the very best we can be, while also cheering on the competition? This is where collaboration comes into play.

Within two departments of the same organization there is never a benefit of one group winning at the expense of the other. Any loss for a department is a loss for the organization, a step backward. With collaboration, we can still push forward, but with the one big goal as our mutual true target. At this point, potentially sacrificing the department's target for the good of the organization becomes an obvious call. If the production department misses the volume target, but the consumer gets the right quality product, and that aligns with the one big goal, then it's the right call. Do we wish the production department had met its target? Of course, but what's done is done. So now we focus on preventing a repeat and on taking the action that drives us toward the one big goal as efficiently as possible.

This idea of collaboration trumping competition is really simple to act on when there's a clear direction. The role of the leadership is to apply this concept in the action alignment process, in order to reduce conflict at the point of execution.

The Final, Approved Master Plan

This document is the business plan. It's the document you will have with you in every meeting and that you will refer to when talking to department heads and frontline employees. You will be aligning others with this document in the daily CEO updates. This is the plan that will form the basis of everything that you and your organization will do. No off-plan activities will be done unless, during one of the review sessions, that activity has, for situational reasons, become a higher priority than one of the items on the business plan's budget list. When that happens, as it might once or twice per year, then another activity inevitably needs to be dropped. Do not increase the overall scope covered by the final, approved master plan unless there is additional budget to fund it.

In my 30 years of production leadership and management, there have been three documents by my side: an action master plan, a KPI scorecard, and an individual work plan for the employee I'm talking with. These three documents drive everything!

Action Items for Step L

The result of Step L (Lead with the Frontline) is having a master plan that is aligned with your direction and the principles for the organization, and that combines the detailed and prioritized action plans of each department. This single master plan document will be a guide for your discussions with department heads and frontline employees. Although it will continually evolve through monthly and quarterly reviews, it will form the basis for further decision-making and resource allocation.

As a guide to creating your own master plan document, complete the exercises below:

1) Answer These Questions

- Who needs to know about your direction, your ground rules, and your plan?
- How will you ensure they all get the message?
- How will you get input from the experts about how to most effectively reach the company's destination?
- How will those experts work with others and communicate about the help and support they need from you?
- What input do you have for the master plan?
- What priority calls do you need to make?
- What is your final master plan?

2) Complete an Alignment Communication Plan

Develop a communication plan for sharing the outcomes of Step A: Aim from the Heart with your organization and inspiring their support. For this exercise, structure your communication plan, and also ensure that there is transparency in the actions to be taken throughout the company to develop the master plan.

- How will you communicate the requirements with the entire organization?
- Who will take each action and by when?

3) Assign Alignment Action Planning

Assign the alignment action planning to each division, along with a timeline for completion.

Starting with the operations and sales groups, assign them the task of developing an action plan for their division, with deep and documented involvement from the frontline employees.

After reviewing those two action plans with your leadership team, align on the priorities and trim down the plan to focus on the priority areas that will have the greatest positive impact on the direction of the company (the one big goal).

Once those action plans have been completed, send them to all the support groups and ensure that each of those groups delivers a detailed action plan to support the sales and operations divisions – again, with the involvement of their frontline employees.

For help with this exercise, you can download a template for the alignment action plan from www.makeitrightbook.com/bonus. This template will guide you through assigning, allowing, and encouraging your best frontline employees, along with a small group of guides, to put together compelling and effective action plans. Although this will not yet be the complete plan – and they should understand that – because it is still to be aligned and prioritized, they will have already started to build ownership. They will already be generating excitement amongst their peers, and will be actively looking for solutions to the most challenging issues.

4) Pull Together the Master Plan

The master plan is an amalgamation of the actions plans, which were created by each division using the same template.

At this stage, your direct leadership team will review the plans with the overall group that was involved in creating the plans, and ensure that they understand and complete a prioritization process for deciding on the final plan. This master plan will then be shared transparently with all divisions to seek final alignment and commitment.

The reasons behind each priority in the plan will be clearly spelled out, and budgets will be aligned with those priorities.

Bonus Materials

Step L is often the step that concerns leaders the most. "Will we really get a benefit from asking the frontline employees to develop the action plans for their department?" "Will the time and effort required to train, educate, and engage those frontline employees really pay off?" "Will allowing the frontline employees to develop the action plans lead to disappointment once the leadership team needs to reprioritize in the master plan?"

The bonus materials below are designed to help you understand why this step is not only good business but is also a very effective method for building ownership, a key factor in the overall success of a business plan.

Direction vs. Instructions

Have you ever been lost? Either literally – in a forest, a new city, a crowd – or figuratively, during a time in your life when you felt you had no idea where to go or what to do? That sinking, empty feeling in the pit of your stomach is a good indicator that your direction has been lost. We sometimes feel like we don't know where to go or even in which direction to turn. If you have a family with you, depending on you, during this time, those feelings can be multiplied.

We can feel like this particularly in the thick of challenges, when things seem to be slipping away from us. At that point, we don't necessarily need instructions to follow; we just need direction. We need to be reminded where we are going, and then we can then determine the critical actions to take to get us there. Although we may need support (more about that later), it is most important to be reminded of where we are headed and that the current struggles are worthwhile in the quest to get there.

Every communication is an opportunity to point toward the direction, and benefits of this should not be underestimated.

Confidence vs. Control

There is a term called *emotional contagion*. Research has identified that the emotions we display are contagious to others. Think of walking down the street and smiling at others, prompting smiles to come to you. Now imagine walking into a room feeling frustrated and angry at the world and the impact that would have on the people there. When you try to control people, they feel that control and are programmed to immediately resist. Alternatively, if you communicate with genuine confidence, that emotion is also spread.

Confidence is a critical trait for creating competence. Confidence gives people the willingness to try, through which they learn and build competence. Indicating that you think the frontline employees are capable of and are most effectively placed to support the company's action plans demonstrates your confidence in the people who make the frontline decisions each day. With every communication there can be a confidence boost for the organization that is a springboard for improved results.

Transparency vs. Restriction

What should we share and what shouldn't we share? A couple of our ground rules will help with this.

We are transparent when whatever we share is, to the best of our knowledge, true. There is no room in your addresses to the workforce for politicking. The thing about facts is that they are facts. They are not opinions, conjectures, or wishful thinking – they simply are. They are not good or bad – they simply are. Facts don't need to be dressed up or dressed down – they are just facts. I often see fear about communication. I see this with people who don't want to share with the boss what they see as bad news. ("She'll be upset and it will spoil the meeting.") I see it with leaders who don't want to share positive news with the employees. ("Maybe they'll expect a pay raise, and I don't want to get their expectations up.") Information is just information. It is what it is! In my experience, the only bad time to share information is when it's too late to do anything about it. (No surprises, please.) Transparency breeds confidence, and we have seen, above, how important confidence is.

Function vs. Fashion

What is the correct attire for a CEO? What about the correct attire to wear when doing a video communication? Should you still do a video communication if you look a bit tired? Will it be okay if your hair is blowing around?

Here's an easy way to think about this. When your sole intention is to help others, it makes absolutely no difference what you're wearing or how your hair looks. Those aspects are only important when the priority is you; if you're more concerned about how you look and what people will think than you are about the message you're sharing. When you focus on helping people with your message, everything related to you will fade away. If you 're walking the factory floor in a hot and humid environment and want to spend time there talking to the frontline employees and seeing how you can best support them, it's unlikely that you'll be able to concentrate while sweating in a suit.

This issue doesn't only come into play when communicating. It is in every aspect of what you do. Clothes do not make the man. Impact does.

Authenticity vs. Image

I love the terms *authenticity* and *vulnerability*. Sometimes people say, "I don't understand why they won't open up to me," or, "They're not the sort of people who share their feelings."

Sharing is a direct reflection of expectations. If you look perfect to your team, they will want to look perfect to you, even if that means hiding the truth. If you're willing to be vulnerable, they will be more willing to share their issues with you.

Relationships are not formed and strengthened during the best times, but when the going gets tough, when the real person emerges, when tempers get frayed and authentic personalities bubble to the surface. We all have the ability to identify a mask, to tell the authentic from the fake. The human face is made up of millions of micro muscles, and each minute movement sends a signal for us to observe. We don't do this consciously; we are far too developed for that. We notice subconsciously.

The only way to avoid being considered fake is to be authentic. This takes guts for most of us, because when we are authentic we let down our guard and we feel that, in that state, we can be most hurt. However, when we are being authentic, that's the only time we cannot be truly hurt. When you are authentic you are simply yourself, so whatever comes from you at that point is you. Although some others might not like it, neither they nor you can deny the authenticity, and that is where strength emanates from. Being yourself cannot be challenged. It simply is and you simply are.

Be willing to show your vulnerabilities, fears, and uncertainties, and be open to listening to those of others. In this mode, you'll be most open to how you can offer help and support so that all move forward.

Listening vs. Talking

You don't learn anything while you're talking. A video address is a one-way communication, as is talking from a stage, and that is fine for communicating the company's direction. Although being face-to-face with individuals is far superior as a forum for listening, there are ways to engage and listen through video. You can provide an opportunity for others to write comments, to send follow-up messages, and to make requests. Those are all good ways to build interaction and, hence, commitment.

During live exchanges, ensure that you have time for others to share their opinions and concerns and to ask questions. Start meaningful conversations in which other parties do 80% of the talking.

There is no need for the leader to know all the answers. A leader who does know all the answers must be leading a very small business. What *is* required is honest and open engagement. Listen, listen, listen!

Union vs. Company

If I am ever faced with a quiet group of frontline employees who seem reticent about sharing their opinions, I bring up the subject of unions. Often, much to the horror of the leadership team, I will ask a leading question like, "What is the point of these unions, anyway?" or make a statement like, "You know, the union will not be able to help you with that."

Unions have, for a long time, been a polarizing force in the manufacturing sector. Often, people love them or hate them. There is rarely indifference, unless the union has very little power.

I believe that unions have served a great purpose in global business and, in some cases, still do. What amazes me is the polarization around this issue. The opposing stances are often depicted as "The unions want the best for the worker, and the management wants the best for the company," and that is where it all breaks down for me. The supposition behind that thinking is that either the worker or the company wins, but not both. In the 21st century, it's only the companies in which the workers win that will be successful over time. In this case, the goals of the union and management should be perfectly aligned, as shown by these questions and answers:

- How do we do what is right for the customer? By having the most capable and successful frontline workers.
- How do we do what is right for the workers? By making them the most capable and successful frontline workers.
- How do we do what is best for the company? By having the most capable and successful frontline workers.

Unionized or not, *there is no excuse for putting any priority ahead of the success of the frontline workers.*

Role Model vs. Figurehead

Whether you like it or not, wherever you go people will be watching. Because you are the CEO, they will be scrutinizing your behavior, either to see how you do it right or to catch you doing it wrong. Either way, they will be forming their own strong opinions and will be sharing those opinions with others. You are and will always be a role model. In this environment there is one sure way to fail, and that is to try and be perfect.

We are not attracted to perfection. We are attracted to authenticity. When we step away from what we think others want to see, what we think others expect, and how we think we should be, then we can be free to be ourselves.

A few tips: Don't set rules for others that you can't or won't follow. Don't set standards for others that you don't really care about. Don't ask a person to do work that doesn't absolutely need to be done.

Video vs. E-mail

An effective way, these days, to send a personal message to the organization, is in a video. Just get started. The first videos you make may be a bit screwy, but that's okay. They may feel a bit forced, but that's ok, too. If you're unsure about how to go about it, you can follow this guideline:

- Ask a rhetorical question. For example, "Do you know which team achieved _____in January?"
- Introduce yourself. "This is David, your CEO, wishing you a great morning."
- Mention the business direction. "As we push towards becoming...."
- Tell them what's important. "We need to remember...."
- Get back to your opening mention of team recognition. "In January, A Team achieved...."
- Give your thanks and ask for continued support. "I appreciate...."
- End with when you will send the next video. "I'll talk to you all again tomorrow."

Each time you make a video, you will feel more comfortable with it. And every day there is something to be proud of, someone to thank, something you have seen that you want to draw attention to, or a message to reinforce. Every day you have an opportunity to whisper in the ear of every employee – take it!

You can find further inspiration and examples of CEO daily messages here: www.makeitrightbook.com/bonus.

Being Inspiring vs. Being Liked

You are not trying to make people like you. Whether they like you or not is dependent on hundreds of factors and is completely out of your control. You are simply there to help them drive the business in the direction you have set. In order to do that, they need to be crystal clear about your direction and they need to know specifically how they are expected to move toward that direction. They need to be reminded and encouraged frequently, and they need a leader who inspires action.

Be the leader you would want to follow, and lead authentically.

Knowledge vs. Experience

Too often, I hear the phrase, "But I did it that way last time and it worked." Although we can learn from experience, it's costly and time-consuming. A far more effective method is to learn through increasing knowledge. If you think training your people is expensive, imagine the cost of having untrained people on the frontline.

I saw a great example of this when I became managing director for a large printing company. In order to learn more about the roles and capabilities of the employees, I was walking the production floor and looking at a huge, new printing line that had been bought from one of the world's best printing machine manufacturers, which was based in Japan. This equipment was state of the art hi-tech and had been purchased for millions of dollars.

As I walked the line with the operating team leader, I asked a few questions. "What is this switch for?"

"Oh, we don't use that function," he said.

"How do you use this computer panel to control the color density?"

"Oh, we don't us that screen. We use the offline system that we also use for the other lines."

Time after time, I asked how the team was using the cutting edge, hi-tech features of that multi-million dollar printing line and the answer was the same: "We don't use that."

It was not really a surprise. The machine manufacturer delivered machine manuals in Japanese and English, but the machine operators were Thai and spoke neither of those languages. The Japanese team that had installed the line brought only one Thai translator, so the training and technology transfer was limited. The line operators had all come from another printing line that was eight years older, and they were experienced printers – they already believed they knew how to run a line and felt that the new features were "not really necessary."

Imagine the advances in technology that happen in eight years. Imagine having an iPhone 7 and using it like the first ever iPhone. Imagine the functionality that would be lost if you never upgraded your skills. Imagine being 18, but with only the knowledge of a 10-year-old.

It often seems that when people join a company their ability to continue learning is severely limited. Either there is no time, no program, no support from the boss, or no culture of learning. I have seen employees berated for asking to go on a course to learn how to use Excel more effectively. The manager looked at the person asking like he was trying to cheat the company and wanted to learn Excel "for his own benefit." When I present it like this, it sounds ridiculous. It *is* ridiculous. If you're not training your employees, allowing them to study, encouraging them to grow and explore their boundaries, then you are cheating them out of having a life and you are cheating your company out of success.

Training must be another non-negotiable. Put the budget aside. Ask your team members what training they need or want in order to become more effective, and let them fly.

One criterion applies: they had better practice and apply that training in their jobs.

The Why vs. The What

When we train employees, the training is often on *what to do*. "To get this ideal result, push this button, pull this lever, clean this, and measure that." We tend to do this in training sessions because time is short, fast results are required, and there is a wide range of starting knowledge in the group being trained. So it feels like there is little time or energy for explaining *why*.

"When I push this button, what actually happens?" "Why does that happen?" "What is the mechanism that allows that to happen?" "If I push it hard or soft will it have the same impact?" "If I push it a bit earlier or a bit later how will it affect the product?" "If I forget to push it what will happen?" "Is there an alarm or a reminder?" "If I push the button late and then do this what will happen?" "Why?"

As you can imagine, there are an infinite number of questions for each instruction about what to do. While the why questions are time-consuming and sometimes frustrating, they are all valid questions. Knowing the why allows the individual using the equipment to strategize, to find new, and perhaps more optimal, ways of working. It allows exploration, growth, learning, and – eventually – mastery. Although all of the whys cannot be addressed during training, there must be a method outside the training room for further exploration. Part of this happens through on-the-job training, but much of it will require additional resources.

Encourage and allow employees to have access to the resources they need in order to continually expand their knowledge. Experience is helpful, at times, but is never a substitute for knowledge, and knowledge requires updates.

Education vs. Entertainment

Many employees commute to work and spend that time listening to something. Audiobooks and podcasts are now becoming very popular.

A huge range of subjects is now covered through those two media. Listening in this way can be like taking a training course while driving a car or sitting on a bus. Once again, the low cost of mobile technology has opened up the option to be in a classroom while in transit. Providing audiobooks or podcasts that are downloadable at work gives employees an additional perk and opportunity to learn.

Your Desire vs. Your "But"

During the action planning process, it's likely that you will hear the comment, "We'd like to do that, but...."

This "but" is a classic form of self-deception. Whenever I hear it, my mind shifts to translate whatever the person is saying to, "I'd like to, but I can't be bothered."

You need to call people out on this excuse to not take action. I mentioned earlier the classic excuse of culture. "But it's not our culture." Well, that's nonsense, and so is "But we've never done it that way" or "But the boss won't allow that" or "But that will cost too much" or "But the company doesn't want to invest."

Challenge everyone on these kinds of assumptions. Companies are desperate to find new, more effective ways of getting things done, and successful companies are willing to invest in solutions that have strong returns on investment (ROI).

Call out the "buts" and gradually get them eliminated.

Taking a Risk vs. Not Trying

Asking the question "What's the worst that can happen if...?" can help with making a decision on whether or not to try a new idea. Rather than saying no to an idea that's different, ask, "What is the worst case scenario if we try this and it fails?" Usually, the worst case is manageable. Asking this question frames the fear and allows us to make conscious decisions. Once we have really faced the fear of what *could* happen, we can then make an informed choice on whether or not to go ahead. If the potential worst case is "not worth it," then we will step back and look for a different option. It's always okay to make the choice not to go ahead. It's not okay to ignore the situation or to not make a conscious choice.

Contracts vs. Ambiguity

It's beneficial to have contracts between departments, in the same way there are contracts between customers and suppliers. The purpose of the contract is not to bring it out at each discussion and use it to hold the other department accountable. There may be times when that's necessary, but the real purpose of a contract is to foster understanding. It is to ease the sometimes difficult discussions that allow groups to figure out how best to work together to achieve a common goal.

The interdepartmental contract covers if-then scenarios that could occur. Let's look at an example between a planning department with a major goal of meeting the sales volume demand with minimum inventory, and the production department with the main goal of producing good quality product at minimum cost. Having very regular product changeovers with a high flexibility of production are conceptually great for the planning department. From their point of view, the production team would make whatever was needed, whenever needed, and on demand. For the production department, that high level of flexibility comes at a high cost. Monitoring and ensuring quality during production changeovers becomes harder to maintain and creates higher scrap costs.

Clearly, there are conflicts of interest involved in interdepartmental goals. We have a company goal to grow the company by producing quality product at the lowest overall cost. Overall cost includes inventory cost and production cost. Therefore, there is a sweet spot between the cost and frequency of changeovers and the cost of holding inventory. The contract between departments is where we discuss this sweet spot.

Both departments understand the challenges and goals of the other and choose to collaborate for the good of the company.

The contract also allows us to discuss ahead of time various scenarios that will challenge the status quo. What if there's a sudden market promotion opportunity that would require a change in the production plan? How will that be handled? What are the criteria we will follow in such a case? What are the guidelines, and how will we reach a decision, even though one group's KPI will be negatively affected? How will this decision be communicated to the leadership so that they understand that the right decision was made? The interdepartmental contract allows the difficult discussions to take place outside of the heat of battle, so that when those situations do occur there is a predetermined course of action to follow and it can thus be carried out with less stress.

As the departments discuss their mutual contracts, there is a trade of giving and receiving. What will each department give to the other (information, support, service) to help them be more effective, and what will it need to get? This exchange of needs between the two departments sets the deliverables that can then be discussed in regular get-togethers. These deliverables are a subset of the contract, and they frame the interdepartmental discussions.

Long Term vs. Spot Buying

An alternative to having a long-term contract with a supplier is to do spot buying. Raw materials and equipment purchasing groups often use this method. Let's say a piece of equipment fails – a gearbox on one of the production lines – and there's not a spare on hand. A typical purchasing process may be to get three quotations, choose the lowest price, and then purchase the equipment. That is the process of spot buying. It's a time-consuming process to follow and requires a lot of documentation, because it involved searching for the best price when the purchase is needed right away.

The alternative is to have a contract with a preferred supplier who will give you a set price for the year. This only requires contract negotiations once per year, rather than on every occasion a part is needed. At the end of the year, the purchases are assessed in light of what could have been bought through spot buys, and the following year's contract is set – either with the same preferred supplier or with a better-value alternative.

In all cases, the purpose of such a contract is to allow the difficult discussions to happen beforehand, so that when a part is needed there's a balance regarding speed and cost. In the same way that discussing interdepartmental issues only when there's a problem is inefficient, spot buying is rarely an efficient way to operate.

Discussing vs. Assuming

Don't let the talk of contracts, negotiated agreements, giving/ receiving, and hard discussions weigh you down. This is not a difficult or complicated process. It's simply a discussion or chat between departments that leads to alignment on what each will do to support the other. While the initial thought about the interaction of each department may take some time, the process will develop. There is no need to get it perfectly right first time. Any progress toward discussions for mutual benefit is positive and should be commended. Year after year, the process will be refined and eventually there will be no need for a written contract. The norms, processes, and systems will be set in a way that fosters collaboration.

For many organizations this process needs a kick-start. Keep the contracts simple, keep them in plain language, and keep them relevant to the goals of the business. Think of the process more as a fireside chat than a negotiation. The attitude is: "We are all here to achieve the same outcome, so what's the best way to get there?"

Our Departments vs. My Department

I have seen departments and groups in many organizations striving for position. Each wants the ear of the CEO and wants to be supported in pursuit of their own objectives. There is often a discussion about resources, usually about how many more resources are needed, and there is often a power play. This is sometimes called *empire building*, where the individual department wants to become recognized or superior to others. In effective companies, "who serves whom" is a moot point. In the end, we all serve the consumer, because she is the only one who pays the company.

I cannot stress enough that the surest way to kill a company is to strive to strengthen one department at the expense of others. This power play deflects all resources from the goal of growing the company, which must be the single focus.

Previously I distinguished a difference between the core departments of production and sales, and the support departments, which is everyone else. All other groups are designed to serve production and sales, but also to push and challenge them. Don't get hung up on who serves whom. Keep going back to the company direction and considering what is best for the whole.

Stability vs. Flexibility

We all want to be infinitely flexible. A company that is flexible and agile will be fast to respond to market demand, will be able to operate with lower inventory, and will be able to innovate more effectively. However, there is a need for stability before flexibility.

Companies can be designed to be agile, but there is always a learning period. During this learning period, there is a need to reduce variability before speed is increased. There is a need to become stable before adding movement. Before gymnasts perform a triple backflip, they master the double. Before children learn to walk, they learn to stand.

It's important for each department to know where the organization is in relation to stability and flexibility. Before requiring three product changeovers per day, the production team needs to be able to perform two with excellence. Excellence in changeovers means rapid execution and vertical start-up, with minimum scrap. If this cannot be achieved at two changeovers per day, then adding a third will cause additional variability, cost, and lost time. This additional waste affects all departments and, consequently, the consumer. During discussions on alignment between departments, there must be a mutual understanding of the current level of capability that makes forming the baseline level of flexibility possible.

Growth vs. Comfort

There is a human instinct to play it safe. This makes good sense when we think about crossing a road or walking close to the edge of a cliff. The issue is the line between safety and security. Safety is necessary for survival. Challenging security – allowing discomfort is necessary to grown. Comfort is the antithesis of growth.

Setting a goal each year of 70% production efficiency and hitting it is comfort. There is no challenge there. Yes, there may be a reward for meeting the target, but where is the growth? Setting a goal of 78% efficiency and hitting 76% means we missed our target, and yet 76% is a lot better than 70%. However, missing a target does come with consequences. If we budgeted on 78%, bought materials for 78%, set sales goals based on 78%, then missed the target, that has real-time business consequences.

So how do we walk the line between safety and security? I recommend applying the concept of setting dual targets.

A *firm goal* is one we absolutely commit to. It is one we strongly believe we will hit if we do the things we commit to do. Of course there are always unforeseen circumstances, but with the best of our ability we believe this target can and will be achieved. The firm goal is used as the basis for budgets, financial planning, material ordering, and sales commitments.

A *stretch goal* is higher. It reflects what we believe could be achieved if several factors go better than predicted. There is a lower level of likelihood that this goal will be achieved, but it is possible. This stretch goal is used primarily to motivate the team to new personal bests, higher levels of achievement, and, therefore, greater rewards. Planning is also done to ensure that the business can take advantage of the stretch goals, should they be achieved. In such scenarios, the departments affected need to have contingency plans. This aspect ties into the if/then scenarios discussed in the interdepartmental contracts and what is given and received between departments. If we achieve 78%, what will need to happen in other departments to take advantage of that great result?

Ownership vs. Accountability

We have talked a lot about action planning and alignment. Let's take a look at key performance indicators (KPIs) in more depth. KPIs are designed to inspire individuals to greater performance. Although they are also an important metric for aligning interdepartmental activities, I see KPIs primarily as a personal inspiration tool. Built up from the firm goals and stretch goals of the frontline employees, departmental KPIs then need to be assessed and interlinked throughout the business. KPIs reflect a commitment from the individuals on how they will add to the value of the business. It is simply unfair to have an employee strive for a goal that has no impact on business results. It's very effective to link each individual directly to business results, and there should be a dollar amount applied to the measurement.

In support departments, how can a dollar amount be applied? Let's take an example of HR and their goal of having an effective pool of talent available for each position that needs to be filled. What is the cost to the business if a position needs to be filled and there is not a suitable candidate? Certainly the company will suffer or there wouldn't be a need for that new person anyway, so what is the financial cost? For a planning employee whose goal is to optimize the production schedule to meet the needs of the sales team while minimizing inventory (there are those simple descriptions again), if the schedule is not met, then there is either a lack of product on the shelves (lost sales) or a reduction in inventory that then has to be made up sometime (additional production cost).

Each employee needs to realize that they were hired for a fee and that the fee must be paid back in the work they do. In other words, the employee must add value to the business. If the individual's value to the business cannot be measured in dollars, then there may not be a need for that individual.

Am I really saying that we have to justify how many dollars we pay our cleaning staff to keep the facility clean? Yes, I am. You have a cleanliness standard for a reason, and that reason better not be just "to look good" (although there is some value in that if you leverage it well). Cleanliness will impact quality, safety, investments in spare parts, investments in refurbishment, and plant utilities costs. If you can't put a cost/value on it, then there will always be discussion about whether the number of cleaning staff is appropriate.

So, yes, there should be a cost and value KPI for every individual, and that individual should clearly understand it.

Home Priorities vs. Work Priorities

I am a big fan of healthy and happy employees. When home life is a mess, work life is a mess. When health is a mess, work is a mess. When people are unhappy they make terrible decisions. Healthy, happy employees are good for business. We therefore need to support the employees to be healthy and happy outside of work.

Although we can't manage the home lives and personal circumstances of individuals, we can certainly set an environment that positively influences them.

We have talked about food and restrooms as having an impact on employees at the facility, but there are more options that can be introduced to help with this issue. In a traditional production environment, flexible working hours can be a real challenge, but do they really need to be? Does the entire team on a production line really need to all change shifts at the same time? Wouldn't it be better from a business and continuity perspective if the team members on a line gradually changed from one shift to another?

In the past, shift patterns were set up with everyone changing shift at the same time, in order to keep control. We want everyone checking in at the same time so we can control who is here on time and who is missing. The traditional shift start times are 7 a.m., 2 p.m., and 10 p.m., and we want everyone to be on time.

But what if, for personal reasons, two members of the morning shift would rather arrive at 9 a.m. and two members of the afternoon shift and night shift would also prefer to start two hours later? Shouldn't they be able to decide for themselves, between themselves? As long as there is no negative impact on the business, on overtime, on transit availability, etc., wouldn't this make a bunch of employees more happy and engaged? With the newer technologies of finger scanning and automatic payroll systems, this is now easier to manage.

If we stay stuck in old paradigms that served an industry fairly well for 100 years, it might be difficult to change. However, the world has shifted. Technology has made it easy to manage hugely variable situations. What is required is for the leadership to be able to see the benefits in these shifts and leverage them to support our people.

There are hundreds of ways to help people with their home lives in order to get them to work happier and healthier. Learning some of the more unique best practices from around the globe, and being mentally flexible enough to assess them, could support a breakthrough for the business.

Prioritization vs. As Much as Possible

Each department has submitted an action plan, and we have looked through the plans for alignment between departments and their associated KPIs. How do we now pare down this list of action plans to decide on the real priorities?

The likelihood is that each department has ranked their action items, which will help. If there is a natural alignment that has occurred during the interdepartmental discussions, that's fantastic. Go with that priority as much as possible. If there are areas where decisions are needed based on resources, budgets, or strategic choices, then there is a tool called a *prioritization matrix* that can be helpful. A template for this method can be found in the downloadable materials at www.makeitrightbook. com/bonus. It ranks the priorities based on a set of important outcomes, and gives each of them a score. The outcomes can be weighted in order to ensure that the correct level of importance is assigned. Once the template is completed, a number will arise for each activity. The higher the number, the higher the priority.

Once priorities have been assigned to every activity, we reassess to ensure that the overall business master plan is congruent throughout.

Saying No vs. Being Nice

There will be items in the department action plans that will not be a priority and that will be dropped after the business master plan has been made congruent throughout. This is a time when departments can become sensitive about having felt ownership for offering up an activity in the action plan that was dropped, and there may be some defensive action to prevent it from being dropped.

The process you have followed up to this point has given you good reasons for the priority choices. Although the action-planning process has been collaborative and has involved a wide section of employees, the decision-making process for the final action plan is the responsibility of the leader. You will have to say no, and you will need to give the rationale for saying no.

Most people want their ideas to be considered. They don't need them to be implemented as long as there has been due consideration. Saying no is the responsibility of the leader and is necessary to keep the business focused.

In many instances, a department will want to have many priorities, to make their department more needed, more impactful, or more important. However, that is often a sign of lack of clarity or of confidence. When there are very few priorities, which is what we're aiming for, it is harder to skip one in order to work on a more attractive one. It is hard to bounce around and avoid making progress. This is why, as leaders, we want only a very few priorities. Progress – or lack of it – becomes very clear.

Full Funds for Priorities vs. Some Funds for Everything

It's nice to say that we will fund everything that has a positive ROI, but this is not possible. There is limited capital to be allocated, so the way to determine which activities get funded is to start with the top priority and assign the money, resources, and time required to make it a success. Assign this priority across all departments that are involved with it. Then move to priority two. At some point, your money, resources, or time will run out. At that point, you make the decision to find additional resources or you reduce your commitments.

Summary of Step L

In Step L – Lead with the Frontline, we complete the communication plan. It is broken down into annual, quarterly, monthly, weekly, and daily segments. There are guidelines in this chapter about what to share and how to schedule the communication planning process. This communication plan will form the backbone of how you share your message, direction, and expectations with the team. It will remind you to encourage and recognize their achievements, and it will position you as a role model for the values of the organization.

We put together the details of the master plan, with input from the frontline employees and departments, and set transparent priorities for where to focus our resources.

This master plan now forms the framework for all actions within the business.

---○---

CHAPTER 5

---○---

I: Inspire with Information

"If you're the village blacksmith and a model T comes along, you better become a mechanic. People's lives are better when they get news online versus having to wait for the morning paper. It's a lot more efficient, a lot more real time, a lot less waste."
Marc Andreessen

Through every stage of our journey to the Caribbean, from inception to completion, we need to know if we are on track. There is no point paying for the accommodation in advance only to find that the trip has to be cancelled because there is an insufficient budget for transportation. Checkpoints, milestones, and indicators of progress are essential. However, taking extra time to collect, analyze, and report data is time wasted. How, then, do we ensure that the right data is in place to allow effective decision making, while spending minimum time preparing this data?

Step I in the ALIGN Process is to Inspire with Information. This chapter explains how to:

- Pull together the right metrics in as little time as possible.
- Ensure that the data is presented automatically, in real time, and in a manner that allows rapid decisions to be made.

What Really Happened?

One of the crucial documents that I believe is essential for the manufacturing CEO is a scorecard. Several years ago, there would be monthly updates to the scorecard with a huge effort-load prior to each update. The data would be reviewed in monthly reviews and decisions made for the future months, based on the data collected. This is simply not acceptable these days.

To start with, there should be no need for taking extra time to prepare data. Every form of data, whether from sales teams, production lines, the planning team, or the quality group, can now be automatically generated in the format required to make decisions. In addition, with the right systems in place, this data can be available in real time, minute by minute, for people with authorized access to use in making immediate decisions. When real-time data is sent directly to the computers, tablets, and mobile phones of employees, it is possible, and expected, for the leaders to know what's happening in each division at all times. This level of transparency regarding the factory and supporting departments gives the facility a cutting edge ability to provide help and support where needed in order to keep the business moving forward.

Transparency is essential for a business, but it is a culture shift that is often difficult for the employees to adjust to, as demonstrated in the story below.

"Show Me the Cell"

I enter the room to find ten employees, from a range of levels, sitting around the conference table. All look slightly apprehensive. The chairman has brought me in to assess the operation and make recommendations on how to improve the profitability of the company. This is a large organization that's attached to a huge conglomerate, so the stakes are high.

In this type of engagement, there is always a level of tension. The manufacturing team doesn't want to expose themselves too much and leave themselves open for a need to make huge changes. They're used to incremental gains being accepted as good enough. 3% per year improvement is considered progress.

As an advisor, I need to strike a balance between fostering cooperation, which can only happen once trust is established, and helping the team see the positive potential in breakthrough performance. Trust is always the key, but we don't have days to build trust, so it needs to come fast.

I share some about my background, and try to establish a bit of a rapport. I inform them that I know what they are going through, and I know I'm an outsider. I confirm that they know their business better than I do, but I also tell them that I have seen a lot of other manufacturing companies, and so there may be some ideas or thoughts I can give to support them.

Then we establish the baseline. Where are we now? What is the real, current performance of this division?

I say "real" because we often live in a world of self-delusion. We like to feel that we're doing the best we can, that obstacles are inevitable, that despite those obstacles we are making progress, and that our progress should be positively recognized and not challenged.

In the meeting, I am presented with the manufacturing results. Typically, there are such items as scrap (product that has to be thrown away as it doesn't meet specifications), efficiency (how effectively does the production line run), quality (typically measured by consumer complaints and some in-process measures).

It all looks quite good:

- Scrap: 3.0% (the world benchmark is about 1%)
- Efficiency: 85% (the world benchmark is about 88%)
- Quality: 50 consumer complaints per month (the world benchmark five complaints per month)

"How do you calculate these numbers?" I ask.

"We use the global standard definitions"

"That's excellent. Can you tell me the definitions?" While it's always difficult to challenge a team, my stance is to "trust but verify." – I trust that you are telling me your truth, but I would like to verify so we know that we both understand the same thing, the same truth.

They show me the definitions they used and we see that we are aligned, that the definitions are the global standard.

I then lead them through one example, the efficiency figure, to further verify.

Their calculation for efficiency is *actual good product produced* per *total product that could be produced in that time period.*

This makes sense. For example, in the time period of 24 hours, with a production line capable of running at 500 pieces per minute, the total possible production would be calculated at 500 pieces times 60 minutes times 24 hours, which means 720,000 pieces. If the team made 720,000 good-quality pieces in this time, they would be able to claim 100%

efficiency. We are all in agreement about measuring efficiency using this calculation.

My follow-up question, however, is greeted with some concern by the manufacturing team. I can see it on their faces and in the way they squirm in their seats when I say, "Using this calculation, 85% efficiency would be 612,000 good-quality pieces per day."

It's difficult to argue with math.

I go on to ask my question. "We don't seem to be making 612,000 pieces per day, so where could the issue be?"

It's then that the explanations start to come out. There is self-delusion, not born out of any malice, as these are great people who are really trying to do a good job every day and make everyone happy. Nevertheless, their "reasons" start to flow:

- "Sometimes we need to run the line slower because of…."
- "Sometimes we need to do maintenance or…."
- "Sometimes scrap is a bit higher because of…."
- "Sometimes…."

"I understand, of course," I said, "these are all things that are required in manufacturing, but these are taken care of in the 15% gap between the claimed 85% efficiency and the 100% perfect run. I am not questioning your actual output, I am wondering how the 85% efficiency figure can be producing less than 612,000 pieces."

The "reasons" continue to flow, but there is now a growing sense of unease in the room. The people not directly impacted by the result can see the issue. How can the claimed 85% efficiency equate to a lower output than expected by the calculation? Once again, it's hard to argue with math.

"Can you show me the Excel sheet you use to calculate these results?" I ask.

A definite wave of concern sweeps through the room. Am I questioning the integrity of the team members? Have they doctored the results for my presentation? Am I really asking to see their 'actual' results sheet? Who am I, an outsider, to question their results, especially in front of their superiors and colleagues?

But it's hard to argue with math, and something is not gelling.

After a little searching, the Excel sheet is pulled up on the projector screen. Indeed, it shows the daily results and, over the period of a month, the bottom line efficiency shows at 85%. There is a massive sigh of relief around the room. The team was correct. There was no intent to mislead. These are good people, and they really are achieving 85% efficiency.

But I am not finished. "Can you click on the calculation cell for efficiency, please?" I ask.

Someone clicks on the result at the bottom of the spreadsheet, the sum of the efficiency numbers, and it shows as "85.045,"

"No," I say, "Please click on the calculation cell, the one that shows *how* the efficiency is calculated. Cell F35 would be fine."

Cell F35 is clicked. The calculation shows as 450x60x22.5, which should give a figure for the actual good product produced per total product that could be produced in this time period.

I compare the spreadsheet calculation to the calculation I previously wrote on the whiteboard and that we agreed was the calculation to use to measure efficiency: 500x60x24, meaning 500 pieces times 60 minutes times 24 hours.

"Your production line speed is 500, so why does your spreadsheet calculation show 450?"

Silence.

"And your working day is 24 hours, so why does your calculation show 22.5 hours?"

Reasons, justifications, explanations start to flow. Self-delusions.

In order to make everyone in the organization feel better about the results, the calculation had been adjusted over time. Gradually the world standard calculation had been modified to a convenient local standard. Pressure had been high to achieve 85% efficiency, and the organization had adapted its calculation to relieve some of the pressure. Slowly but surely, the "reasons" had become justified and accepted, or at least gone unchallenged.

Using the world standard calculation of 500 pieces and 24 hours in the day, their efficiency was actually 71.7%.

It's important to recognize that *this discussion did not in any way change the results of the business*. The team was still making about 516,000 good pieces of product per day. That had not changed. All that had changed was our perception of what could be achieved. In one simple discussion, we had identified a 13.3% gap between where the production actually was and where they'd thought it was. That gap is pure opportunity.

There was no need for the team to feel uncomfortable. Discomfort is not a good motivator – but discontent with the status quo is. With that one discovery, we had stripped away the self-delusion. We had shown that there was now an actual 16.3% gap between the current results and being a world-class production line. That was an opportunity worth millions of dollars. With a focused action plan, we could recoup those savings.

To move forward on a journey, we need to know where we are starting. We need to be brave enough to *truly* assess where we are and to maintain the foundations and the direction as we move ahead. Transparency in the data and an ability to look at the real situation in real time is essential for assessing progress.

It's natural to want to move faster, but it's hard to argue with the math!

The Importance of Data

Data is power. In negotiations, in engineering, in finance – in any aspect of life, the person with the best data wins. Accurate data is the basis of any troubleshooting process, whether it's about why you arrived on vacation late or why your bank balance wasn't enough to fund the vacation. Without data, we are blind.

What is the best kind of data? Have you ever been in a monthly meeting and heard a discussion that went something like this when the monthly results graph was shown?:

- "What happened on 14th? These result looks very low."
- "Um, I think that was size-change day."
- "What was the issue? It still looks low, even for a size change."

"Um, I can't remember, but I believe the _____ unit failed. I'll check and get back to you."

The reason for frustrating discussions like this is that the data was compiled just before the meeting and just for the meeting. *Data is not for meetings. Data is for making minute-to-minute decisions. Data is for immediate action.* You cannot fix an issue 24 days later. The appropriate time for making an adjustment has already passed.

Therefore, the best kind of data is real-time data: the type of data that is directly available to the person who needs to make the decision in real time; the type of data that does not need time to be collected, compiled, or analyzed; the type of data that is fed to the fingertips of the decision-makers in real time in a format they can use immediately. Welcome to the new real world!

Only a few years ago, this kind of data was hard to access. Excel was the king of data manipulation, and data collection was still a highly manual process. People would write numbers down on a notepad, transfer them to Excel later when they were at the computer, set up the data presentation to produce the most relevant graph, and then use that for the meeting. If a different type of data or format was required, the individual preparing the data would need to design a new template, perhaps redo some statistical analysis, and then print out the new graph.

Real-Time data

There is no longer a need to do any of those old data-processing steps. In most cases, data can be pulled directly from the source, sorted automatically, analyzed based on pre-programed algorithms, and fed back to the individual's fingertips via a mobile phone or tablet in real time. Yes, exactly at the same time that the data is being collected, it is analyzed and formatted for use. In many cases, a required action based on that data can be automatically recommended, or even automatically undertaken.

Preparing data for meetings is a thing of the past. The data for daily operations can be automatically collated, summarized, and presented in advance of the review. People who need to know can access real-time data updates at any time. With authorized access, there is transparency into the workings of every part of the organization.

This is the new expectation around data. The time taken to collect data should be eliminated; the delay in getting data should be eliminated; and the transparency into the operation should be absolute for those people authorized.

No Discussion Needed

The need for data is indisputable. How do you argue with data unless you have better data? If I know my kid missed arriving home by curfew time by 27 minutes, and I know that because the data for the finger scan required to open the door shows it, that's very hard to dispute. The *reason* for the delay might need to be discussed, but the facts are clear.

When there is real-time data for the operation of a production line; clarity about when the line stopped, started, and why; the actual production speed; the number of rejects and the cause of the rejects, then there is very little discussion needed between the maintenance team and the operating group. There is no need to dispute facts. All the energy can go into finding and fixing the cause of the problem, and this is what we want our teams to focus on.

The Fear of Transparency

A big fear, one that comes up again and again, is the fear of transparency. I do understand this fear, and there's usually evidence in an organization to warrant a certain level of fear of transparency. Quite often in such cases, it turns out that, in the past, there was an ineffective manager who thought that the right response when presented with facts was to shout and moan and berate and criticize and find some way to discipline the person who presented the real data. Why was that manager ineffective? Because rather than embracing the truth, thanking the individual for being transparent, and looking for a solution, that manager closed down almost any opportunity for improvement and, in the process, encouraged everyone around to hide the facts. Many companies have functioned like this for decades, and those companies have struggled.

Now there is a dilemma. Real-time, transparent data is going to expose the hidden details, and there are going to be shocks. I've seen factory managers who were convinced their operations were running at a 2% scrap level find out that the truth was 5%. In a manufacturing company, 3% scrap is 3% of the total material costs plus all of the energy, manpower, etc. to manufacture that 3%. In other words, these expenses come straight out of the bottom line. So that extra 3% of surprise scrap can mean millions of dollars of losses – or it can mean opportunity.

Just because the data now tells you there is 5% scrap instead of the previously thought 2% scrap, *the reality of the situation has not changed.* It still is what it is. The only thing that has changed is that now you know about it.

There are two ways to look at and deal with these newfound losses. You can treat these losses as:

1. **Something to beat people up for.** You can lay blame and have all the other negative reactions that will immediately come to mind in the moment. Someone in the organization has been reporting incorrectly, and they can be fired, or an amnesty can be given. I have seen both approaches, and the amnesty works well if very clear expectations are set around what will happen in the future.

2. **As hidden treasure.** You now have 3% more opportunity to improve the bottom line than you knew about. You now have an opportunity to clean everything up, learn more about your organization than ever before, and find out why the culture thought that hiding losses was the right thing to do.

One company I worked with was so fearful of revealing the real data that they asked me to phase in the real data over a six-month period so that there would not be a big shock that the CEO would notice. I managed to talk them out of that approach, and they were finally brave enough to tell the CEO that the new system they were putting into place would show some big opportunities. No one was fired because of that approach, and they got started looking for the root causes of the issues – and getting great results. When the real-time data system showed the rapid improvement the team had made in a little over two months, they became afraid again and asked me if we could spread the improvements out over a longer period so that the CEO would not ask questions about why they hadn't made the improvements sooner.

When you delude yourself with incorrect data, it is very hard to make improvements. When the data is clear, the solutions also become clear.

When there is fear around transparency in the operations, it is the role of the leader to understand this, acknowledge it, and move forward with whatever actions will have the biggest positive effect on the business growth. Be brave, face your own fears, and build an environment where data transparency is expected and accepted.

Action Items for Step I

It should be clear by now that the automatic, real-time availability of data is an essential element for running a successful manufacturing business, and that ensuring transparency in the data will most effectively identify areas for improvement and support the effectiveness of improvements made. To have the right data for running the business, start by answering the following questions:

- What are the key metrics you will use for tracking the progress of the business?
- How do these metrics support the company's direction?
- How will you collect this data without using additional employee time to do so?
- How will you structure this information so that decisions can be easily and directly made from it?
- How will you ensure that the data is available in real time?

Once you have set clear expectations and requirements, you may need to charter a project to get the data delivered in the most effective manner. Although this will take time, effort, and cost, it will be worth it. There is no substitute for data transparency. Data is power, and so it must be a priority for the business.

Bonus Materials

These subjects of real-time data, information availability, and transparency always raise many questions. There are simply a lot of cultural, political, and fear-based barriers to break down, and that can bring out the worst in people. A transformational leader, one who is willing to lean into these fears in order to deliver breakthrough results, is required. Below is bonus material for consideration as you approach this challenging topic and form your own principles around data management.

Essential vs. Nice to Have

Since there is so much data that could be collected, where do we focus? How do we know we are collecting and responding to the right data? The key is to work back from the action plan. If a critical action in the plan is to improve the uptime or efficiency of the equipment, we will want to collect real-time data on the causes of line stops, the reason for the length of downtimes, the process variability that causes equipment jams, etc. If the action step in the action plan is reducing quality defects, then we need to know the most frequent and critical issues, their root causes, and the data needed to effectively troubleshoot. For a cost-reduction action, we need data on loss analysis, we need frequent tracking on spending, and we need daily tracking of cost vs. budget.

All of the data should be aligned to the information needed to make smart decisions and to know if our decisions are taking us in the chosen company direction on which the action plan was based.

I want to make a quick mention of data for reviews, which I'll cover in depth in a later section. Reviews should use the same data formats that are used in the daily business. There should be no need for additional graphs, charts, or data manipulation for reviews. If you're not using the data to run the daily business, then why would you want to use it to make a decision in a meeting? If you show data in a meeting that should be used in the daily business but isn't, then get that data feed into the daily business operations so the frontline decision-makers can do their best possible work to move the company in the right direction.

Data for Decisions vs. Data for Display

Automatic sorting, analysis, and presentation of data allows data to be presented in a huge number of formats. My only criterion for how the data should be displayed is that it be done in the most effective manner for making a decision. At times, a simple Pareto chart will be enough. At other times, a heat map, a graph showing correlation of several different variables, or a run chart would be more effective. This is where the frontline and engineering teams can collaborate to work out how to present the data in the best format.

Flow vs. Activity

There has been a huge amount of investigations done around describing the flow state. One of the preeminent researchers in this field is Mihaly Csikszentmihalyi, who recognized and named the psychological concept of flow (Csikszentmihalyi, 1990). *Flow* is the state where we feel at one with the task we are doing. We feel like time has disappeared, but we've been deeply focused and effective. *Flow* is the state that all sportspeople seek, because when we are in that state, everything seems to work for us. We are perfectly attuned and aligned with the purpose of our activity. This is the state I seek when I work, and the state I strive to help others achieve. It is when we are in a state of flow that we are truly most effective.

A critical factor of reaching flow is getting instant feedback. We want to ensure that what we are doing is generating the desired result. With our senses heightened when in a state of flow, we need data to pinpoint our direction. Real-time data supports a state of flow. Whether we are on a production line, in a sales call, in a planning control room, or in a quality lab, if we can see real-time data it gives us immediate feedback on which to base our next decision in the moment. In this way, we are *alive* as we make decisions that have a direct effect, and we see immediate feedback data so we know whether our decision has had a positive impact. Real-time data feedback is critical for enabling this flow state in business, and should be the basis of each business decision made.

Summary of Step I

In Step I we have established the importance of having the right data, and we've looked at how this data is most effective when it's available in real time. We have determined that the information, as much as possible, should be available without spending additional time to collect or analyze it, and should be presented in the most impactful manner for making immediate decisions. In addition, we have examined how having the right data can be the most effective tool for feedback and for self-motivating a team.

CHAPTER 6

G: Give Help and Support

"A good leader can engage in a debate frankly and thoroughly, knowing that at the end he and the other side must be closer, and thus emerge stronger. You don't have that idea when you are arrogant, superficial, and uninformed."
Nelson Mandela

For our Caribbean trip, it will be essential to ensure that everything is on track and to occasionally make critical, cross-functional decisions. These reviews between the individuals affected are required to make the decisions.

The word *meeting* fills me with dread, and for good reason. Meetings have become one of the greatest time wasters in business, possibly only beaten by cc'd emails and global conference calls. Yes, it's essential for people to meet with each other and to build relationships, but I believe it is even more essential to share data in a manner that encourages decision-making, and to communicate the direction and goals of the business. A meeting room filled with the wrong people, without a clear agenda, and without knowing the absolute decisions that need to be made is the wrong venue.

Step G in the ALIGN Process is Give Help and Support. This chapter covers the real purpose of meetings (spoiler: it's to make decisions) and how they should only involve the people required to make the decisions, and only last the length of time it takes to make those decisions. Meetings are not for general communication, training, praising, or reprimanding. They are not for relationship building, drinking coffee, or eating lunch. There are separate times and venues for all of those aspects (see other chapters). Meetings (also sometimes called reviews) are for making decisions. Period!

What's This Meeting About, Anyway?

The sole purpose of any meeting is to assign help, as required, to keep the business moving forward at the optimum pace. That help may be in the form of a priority call, a resource allocation, or support in overcoming an obstacle. Either way, the only people needed in the review are those who can listen with the intent to help and with the authority to make that decision. Most often, that decision can come in a one-on-one meeting.

There are several resources about how to hold effective reviews, but I recommend using one guiding principle: *If there is not a clear decision to be made, there is no need for a review.*

In a dynamic manufacturing company, the high number of variables leads to a need for real-time data and frequent decision making. As much as possible, we want the decisions to be made immediately by the people closest to the situation. Occasionally, though, there will be a need to ask for help and support in making a decision or assigning resources. A circumstance requiring a huge number of variables is in the launch of a new product or initiative. The story below illustrates the challenges and crucial meeting required to deliver effective initiatives.

Find the Root Cause

A client asked me, "Why do our new product initiatives always seem to fail?" That was a very honest and vulnerable question. The client had launched a series of initiatives over the past few years and, without exception, there had been a major issue. When they managed to plug one hole, like running out of current inventory before the new product was available, they would stumble upon another, in the form of product released to market that didn't please the consumers. As we looked together through the catalogue of failures, there were a couple of common themes that ran throughout:

1. At each stage of the project there were no success criteria, review processes, or sign off protocols. The project simply rolled from department to department, with each being expected to make up the ground lost by the previous department.
2. There were very few interdepartmental reviews in which check-ins and critical decisions were made about the status of the project. Everyone seemed too busy to drop their daily business and spend time on the new initiative. As often happens, daily business always seems urgent, but in this case, the lack of scheduled reviews allowed urgent to trump important!

Many of the criteria we've covered in this book in the earlier steps come into action here. Without clearly defined boundaries between departments, it's difficult to know who is responsible for what and what criteria are required to be met before passing the project on. Without giving and receiving between departments, it is difficult to know who should do what and by when. Without a common direction, it is difficult to know when the project should be halted or pushed forward with an aligned decision.

My client was struggling with all of these issues. Rather than plugging the gaps with one or two actions, we transformed his organization. We needed to get to the root cause of the issues and fix them once and for all. It took a year to fix the basics and launch one near-perfect new initiative, but by then the process was in place and the operating paradigm had changed. There were no more excuses.

Have Intentional Reviews

The word *meeting* has been so heavily abused that I prefer *review*, but the word used changes nothing. Whether you call it a *meeting* or a *review*, it has one purpose, and that is to reach a decision. Yes, training is important, information sharing is important, drinking coffee is important – but please don't mistake those activities for a decision-making review.

The purpose of a review is to assess and adjust, to make decisions about how to proceed in the most effective manner, and to decide what to do and what not to do in order to move the business in the chosen direction as effectively as possible.

Before scheduling any review, be clear about what decisions need to be made during it, and state that to the attendees. This in itself will eliminate 50% of the meetings. Be clear about who is needed in order to make the relevant decisions, and when they need to attend the review, because they only need to be there for the time needed to make that decision. This will reduce 70% of the people currently in meetings. Ensure that there is an agenda for each scheduled review, so that the use of everyone's time is efficient and so that those who have a tendency to waffle are kept on track. Ensure that the data required to make the decisions needed is taken into the meeting in a manner that can be used during the meeting to make the decision.

Follow those steps and the majority of decisions will be able to be made even before the review, or will be simply a formality during the review.

Reviews are critical for assessments and decisions regarding actions. Do not allow your team's most valuable resource – time – to be eaten up in preparation for or during an ineffective review.

The Importance of One-on-One Discussions

The one-on-one discussion is the foundation for any lasting relationship. On a one-on-one basis we get to know the individual, and can better allow ourselves to be vulnerable and dig deep into the challenges. We can assess where the individual most needs help and seek clarification. Since relationships are the basis of work between people, the one-on-one discussion is a foundation for a healthy business.

For a one-on-one discussion to be successful, both participants need to be fully present and attentive. This means putting normal work and social media distractions aside and spending time leaning into the conversation. For an effective one-on-one discussion, have a clear theme, so that each participant can be asking, "How can I help?" The setting of this theme is primarily the responsibility of the leader of the discussion, but the helping works both ways: How do we help each other accelerate the business in the desired direction?

There is a skill to holding effective one-on-one discussions, and it can be summed up in one word: *Listen*. You don't learn anything while you're talking, and your role as the leader is to learn how to best help and support the other. You also do no one any favors by taking on responsibility or accountability that the other should have (refer to the section in this book about stealing ownership). The skill comes in aligning with how, in your realm of responsibility, you can best support, and with seeking clarification that the support you offer is that which is needed. Then there is decision and a commitment, just like in any review.

Ask for Support

I frequently hear, "We asked for _____, but the answer was no." When I ask why the answer was no, there is a flurry of further answers having to do with things like no money, not a priority, maybe next year, further thinking is needed, and "Okay, I'll pass the request up the line, and we'll see what happens."

We can learn to structure requests for support, so that the answer we get is most often, "Yes, of course." In a nutshell, it is the requester's responsibility to do the groundwork to get the decision-maker to a point of making a positive conclusion and supporting what is being asked for.

When I was doing a factory audit, I noticed that many of the electrical panels had been left open. In a factory environment, electrical panels contain a lot of sensitive electronic equipment and should always be left closed, to protect people from electrical hazards, so that the temperature of the equipment can be regulated, and so the equipment is kept free of dust. Keeping the cabinets open is a recipe for failure.

"Why are these doors kept open?" I asked.

"The air conditioning in the room doesn't work, and if the doors are closed the equipment gets too hot."

I said, "Even with the air conditioning failure, the individual fans in the electrical cabinets should keep the equipment cool until you can get the air conditioning fixed"

"Oh, those fans failed."

"Why?"

"Because the filters were full of dust and so the fans were always overworked."

"Well, when will these air conditioners and fans be fixed?" I was getting curious about how much responsibility was going to be handed over to someone else.

"We keep asking for budget for this, but it keeps being pushed down the priority list. It's been this way for two years!"

Now let's look at this situation again. The fans failed because there was too much dust in the room, and so the filters became blocked. Then the room's air conditioning failed. Now the electrical cabinet doors are open to the hot, dusty room, which creates a much higher chance of electrical equipment failure. And the situation wasn't going to be fixed because it wasn't being considered a priority.

Guess what the number one cause of production failures was in that hundred-million-dollar business? Yes, that's right: electrical failures. Those electrical failures were causing the business *millions* of dollars per year. To fix it would cost a few thousand.

Whose fault was it? The leadership should have identified the issue, but if the maintenance team had known how to ask for help, things would never have reached the situation I saw during my audit.

The thought process for finding a solution looks something like this:

- Equipment downtime is our biggest issue, from a cost and capacity perspective.
- Electrical failures are the biggest cause of equipment downtime.
- Electrical failures are caused by dust and heat.
- Our electrical cabinets are dusty and hot because we leave the doors open.

- We leave the doors open because the air conditioning and fans are broken.

Isn't it obvious where the company should be spending money? Do you see how they could calculate the return on the investment of fixing the air conditioning? Root-cause analysis and data-based decision-making give our people the right tools to make effective requests for support, because it allows them to ask in a way that points out impacts to the company's bottom line.

Be Careful What Message You Send

Let's take a deeper look at the example above. Why did the request for fixing the air conditioners keep getting rejected, and why did the maintenance team *allow* their request to fall to the bottom of the priority pile?

Yes, they knew that the broken air conditioners were making the electrical issues worse. However, a couple of years before, the business had gone through a tough financial patch and had significantly reduced spending. The message filtering down from leadership was that belts needed to be tightened, and all non-essential spending should be stopped.

It was around that time that the maintenance team asked for money to fix the air conditioners for the electrical room. Not understanding how critical heat build-up was as a factor in electronic equipment failure, the leadership saw the request for air conditioning without a context. Not knowing that, without air conditioning, the electrical cabinet doors would be left open to dust, and not knowing that dust was a major contributor to electronic failures, the leadership rejected the request for air conditioning.

The leadership kept on signing checks for repairs of electrical components, without knowing about or fixing the root cause of those costs. Someone on the maintenance team told me that, "We asked a few more times, but then stopped, because it was clear it wouldn't be funded."

If you knew this full story and were that company's owner, would you spend money on the air-conditioning? If you were a maintenance person who was also one of the business owners, would you have made your case more clearly? I'm sure the answer to both is yes. How, then, do we develop ownership behavior to the point that each owner/employee will fight for what is right for the company? How do we give the employees the skills, knowledge, and confidence to deliver what is right for the company? The answer is *training* and *transparency*. If the ROI on spending is positive, it is always going to be of interest to the business owner, whether that is the leadership or a team that has some measure of ownership and pride in the company.

How to Say No

Knowing when and how to say no is just as important as knowing how to ask for support and giving support where needed. Here are some of the times to consider saying no:

- **When the request cannot be linked to any of the priority actions on the action plan.** There was a good reason you spent time making the priority decisions. Staying focused keeps everything simple. While the proposed project may have a positive ROI, the distraction from the priority actions will prevent those higher priority projects from going ahead smoothly.
- **When the ROI is negative.** It's difficult to justify an expense without a return, but make sure you're not only looking at direct, short-term returns. Training is a good example of this. Training should be generating a result that adds more

value to the business than the cost of the training, so take the full value of the training into account by assessing how it benefits the company over time.

- **When resources have already been fully allocated to higher priority actions.** A project might be on the priority list, but you have run out of people, money, or time to get it done. Fully support projects higher on the priority list before you start another that's a lesser priority. In this way, although you may be missing an opportunity, you are spending your resources where they will have the greatest return on your chosen direction for the company.

Saying no should be an easy decision when there is a clear direction and transparency. There may be some disappointment, but if the rationale is clear that your no best supports the business, then the decision should be understood and respected. Most important is to have the conversation about it, listen carefully, then give the rationale, and ask for understanding and support on the chosen projects. The person pushing for the specific project you've said no to has a passion, and your role as a leader is to channel that energy and passion in the needed direction.

The S&OP Review

The Sales and Operations Planning (S&OP) review has special significance in a manufacturing company and, when performed correctly, becomes the heartbeat of the organization. The S&OP review pulls together the two core departments – production and sales – along with all the support groups. The purpose of the S&OP review is to make decisions to ensure that the business is supplied with the right product at the right time to best meet the demands of the customer/consumer. The story at the beginning of this chapter about the failed initiatives was a case in which the S&OP process had clearly failed, and the consumer was being affected by new initiatives that created issues.

A simple description of the S&OP review is that it looks at the demand from the consumer and translates that into plans for production or operations that will support this demand. In reality, there are many different moving parts involved, and so the process of creating and updating the S&OP review needs to be highly integrated, regarding ongoing business, and especially during a time of change. Every department is critical in this review process, and prior to the S&OP review every department needs to be prepared for it.

Let's look at a simple scenario. The customer (a supermarket) has decided to do an unplanned and unannounced promotion to sell our product at 20% off. There is no clause in our customer agreement to prevent this. So, all of a sudden, our product is flying off of the shelves, and they are almost out of stock. They are a major retailer of our product, so we're eager to address this shortfall.

Because the planning team had been doing a great job of minimizing inventory, there is not a sufficient buffer for this unplanned scenario. To refill the shelves and the buffer inventory, many actions need to be taken, including:

- The production plan needs to be changed.
- A priority call needs to be made regarding what production will be delayed (thus minimizing another inventory buffer) in order to fulfill this one.
- In the factory, the production operation is asked to do an unscheduled size change, to make the product needed by the supermarket out of turn, thus going against the pre-agreed production pattern.

- Due to this sudden change the production and maintenance teams are not fully prepared, and so the size change takes longer than normal. Capacity is lost and scrap cost increases.
- To make up the production that was bumped for this sudden need, the operations team works Sunday on overtime, and HR needs to schedule transportation, food, etc.
- The quality department can't encourage enough people to work on a Sunday, so the operation department decides to have one of the line team members stand in for the quality team.
- The maintenance team, rather than coming in on overtime, is asked to be on standby at home.

Although the company was able to make enough product to minimize the time the product was off the supermarket shelves, and sales increased slightly, the cost to the organization was huge. Was it worth it? Did they do the right thing?

Decisions like these are made using the S&OP process.

We always want to minimize or eliminate surprises, but – as we saw in the example above – such a small decision by a customer can create a whiplash effect through the entire supply chain.

Although the S&OP review is not set up to deal with emergencies – because it is a planned and scheduled review that acts as a heartbeat for the business – it is there to assess such situations as described above, and to help with making decisions about how such scenarios will be handled in the future. All departments are involved, as the planning process and the implications truly affect everyone.

Many of the issues discussed earlier regarding meetings are critical here, too. The purpose of the discussion and use of the S&OP process is to *avoid* surprises, assess what-if scenarios, understand the ROI of a proposition, and make an aligned choice. A choice that, even if hotly debated in the review, will then be supported fully by all to give the company the maximum chance of success.

There's enough to say about the S&OP process and review that it could be a book in itself, but what is important about it is that it's an *orchestrated review*, involving all departments, for making decisions about how to best meet the direction of the business. There are software models that can help with the data flow and decision-making process, but the decisions must ultimately be made by humans.

It may seem obvious that all departments should discuss the S&OP process together on a regular and frequent basis to decide on the best courses of action, but the majority of manufacturing companies I visit still do not do this, and those that do could usually do it better.

Attend the Planned Reviews

Assuming you have a planned weekly review where the team gets together, looks at what went well last week, reviews challenges, and makes some decisions about the upcoming week, if the previous week was quiet and it seems like there's not much to decide, do we still meet?

My thought around this is don't plan reviews you don't need in order to run the business effectively, but do plan reviews often enough to keep things on track.

When you do plan a review, then, ensure that it is attended. It's always easy to find something in the day-to-day work that encourages you to drop the scheduled review, but you set it up for a reason. It does not have to be a fixed length, but the act of meeting, reviewing, and making decisions on a frequent basis is what keeps the business on track and going in the desired direction. Don't skip the planned reviews. If, after a time of stability, the review frequency can be reduced, then make that decision. If the weekly review is really not adding value, then cancel it, but don't start skipping ones that are scheduled. Don't let your standards drop, because that will indicate to others that the reviews are not important.

· Action Items for Step G

This section is all about giving help and support. The vehicle for much of this help and support is reviews (meetings). During these planned reviews, the requester of support can share their challenge and ask for the help they need. If their request for support is well-structured and aligned with the company's priorities, the answer will most often be yes.

In planning reviews for your business, consider the following questions:

- What reviews are essential to keep the business on track and make critical decisions?
- What are the specific outcomes expected from each review?
- Who is essential for each review to make critical decisions?
- What is the ideal frequency for these reviews?
- How do you ensure that there are one-on-one discussions for each person in the organization in order for them to be listened to and to receive offers of help and support?
- What is a communication plan that will work at every level?
- What is the structure for each review session?

You can use the ALIGN Process for reviews as a guide for structuring effective reviews and ensuring that attendees are well-prepared for the review to be most successful.

Bonus Materials

Not all reviews are set in an office. In fact, many reviews can most effectively be held directly at the workplace floor. Do not, however, confuse on-the-floor reviews with ad-hoc interruptions of people while they're doing their daily work. All reviews should be planned and well prepared for, or they are distractions.

The bonus materials below cover aspects of reviews that can challenge manufacturing leaders.

Factory Floor vs. Office

A factory or manufacturing facility is a unique place. Most commonly, a high number of frontline people are making a product, led and supported by a few that are office-based. On each factory visit you make, there will be a few decisions that need to be made, and they make up the review portion of your visit. Those reviews may be best done in the office, but that time should be minimized, following all the review guidance above.

Factory visits are great opportunities for you whether you are the CEO visiting once a month or the operations team leader going onto the production floor every day. Your time on the factory floor can be used to share the company direction, model the standards, observe situations, get feedback from the frontline, listen, and inspire.

My preference is to schedule the factory floor visit before the in-office reviews, and to set a good chunk of time aside for it. The purpose of the visit to the factory floor is to offer help and support, assess the reality of the situation, and model the company principles.

Here are a few tips for visiting the factory floor:

- Start at the furthest corner of the factory, away from the offices. This is the area that is least visited by factory team members, and so it's often forgotten. Starting there ensures that you will see the entire factory as you make your way back to the office, and it sends a message that all parts of the factory floor are equally important.

- Always check the quality of the product currently being made, as this shows your commitment to quality.
- Engage the frontline workers by asking questions related to their area of ownership, and allow them to answer (not the leaders). Listen intently, and ask if there's anything you can do to help. The discussion doesn't have to be about the job; personal discussions are often more valuable.
- If an individual expresses a desire to talk more or to share more, schedule a one-on-one discussion.
- Point out anything on the factory floor that you don't understand or that looks different from your expectations. Ask the questions that are on your mind.
- Thank people for their contributions to the company. Let them know how their specific role adds value and is critical to the company's success. (If you don't know this, then make it a priority to find out!)
- Ask people to tell you one amazing thing about what they do, so you can learn something that you didn't already know or wouldn't otherwise have learned.
- Eat in the staff restaurant.
- Visit the factory floor washrooms.
- Follow the same guidelines as above as you visit the office, and engage with any office staff who don't usually participate in regular reviews.

This factory visit, like any review, is scheduled, planned, and expected. It is not management by wandering around (MBWA), which is basically an interruption.

Your factory visit can be a huge success and should be thoroughly enjoyable.

Embrace vs. Resist Audits

There will be lots of requests and demands for audits of your business from customers, authorities, certifying bodies, safety officers, financial auditors, etc. There are two possible approaches to audits: resist them or embrace them.

Resistance is tiring and stressful. When resisting, there is this sense of doing and showing the minimum required to pass the audit, and spending as little time as possible on it. Some audits can be a distraction, but that should only be the case if the audit was unexpected.

I prefer to embrace audits. Allow the experts in their particular areas to point out what you could improve, and the areas around which your team can learn or develop. Ask for input about the latest industry developments, benchmarks, and best practices that the auditors have seen elsewhere. Ask the auditors to identify areas in your business where people are doing a great job and so should be recognized. Use the audit to move further in the direction you want to be headed.

Companies that have higher standards than the auditing groups have no fear of audits and, in fact, use audits to validate their progress.

Choose to embrace audits and amaze the auditors with your openness.

Contributors vs. Suppliers

It is highly likely that the suppliers of materials, equipment, and services to your business know more about their products than you or your team does. A gearbox supplier is an expert in gearboxes. They know the tolerances, the best maintenance techniques, the ideal running conditions, how to check for abnormalities, and more. They know the new innovations that are coming and which ones might be most relevant for your business. Suppliers are a goldmine of training and information for your business. And because they are experts on their topic, they usually love to share this information. Since you are a paying customer and they want to build a good relationship with you, they will quite often share information at no cost.

Treating your suppliers as contributors to the success of your business is great for them and for you. As they help build the capabilities of your team for your business, your employees will be stimulated and exhibit more ownership. The supplier will benefit by having a stronger, growing customer.

Make maximum use of your suppliers to get the most from their knowledge. Allow the relevant members of your team to visit supplier factories, learn about the causes of variability, and determine how best to integrate that equipment, those materials, or those services into your business.

Three Documents vs. Many Documents

During my thirty years in manufacturing management, I came to rely on three documents to deliver on results. The three documents are:

1. **Action Plan** – This includes a statement of where we are headed, the route we will use to get there, and how we will make the journey.
2. **Scorecard** – This tells us where we are, how far we have gone, whether we are still on track, if we need a course change, and how far we still have to go.
3. **One-on-one review sheet for key people** – This is possibly the most crucial tool of all. For each of my direct employees, I keep a brief record of our discussions, the progress they are making, the help that was requested, and what they plan to deliver. That record is used when we review. My

directly reporting lieutenants bring their own action plan and scorecard to our reviews.

Does focusing on three documents sound too simple? They answer the important questions: Where we are going? How do we plan to get there? Where we are now? What help do we need to make progress?

These three documents go with me everywhere. Over the years, I have modified the format and the method I use to optimize them, and I continue to simplify. However, the basic purpose of using the three documents has remained, and they served me very well.

Summary of Step G

In step G we complete a review calendar. We assessed the need for, purpose of, and optimum frequency for each review. Whether a review is a one-one-one discussion or a complex S&OP review, the process needs to be defined, and the specific outcomes desired need to be clarified.

Calculating an ROI for time spent in reviews can be a challenge, but they support the certainty that business depends on people and on smart decisions. The review is where these two elements come together. Time is the most valuable resource. Reviews should include only those people whose time is required to make the decision in the shortest possible time.

For a leader, the agenda for every review is "How can I help?"

CHAPTER 7

N: Nurture Feedback and Recognition

"The more you praise and celebrate your life,
the more there is in life to celebrate."
Oprah Winfrey

It would be unusual to finish our Caribbean vacation without having taken any fun pictures, without celebration dinners, or without a toast or two. It would likely feel uncomfortable not to recognize the contributions of those who pulled the trip together and those who made it a success.

It can be demoralizing to get no reaction and no feedback for our efforts. And few things are more rewarding than recognition. When we catch people doing things right, acknowledge their contribution, and cheer them along the way, they will be more likely to want to continue doing well.

We often get caught up in the busyness of business, with the next challenge that's immediately on the horizon, and we forget to stop and recognize the progress that's been made.

Step N in the ALIGN Process is Nurture Feedback and Recognition. This chapter explains the importance of celebrating the milestones along the journey. It's about how to catch people doing things right and encourage further progress. We'll also discuss how to approach situations that don't go as planned, and how to set people back in the right direction if they go off-track.

What Are You So Happy About?

As the CEO, you do not do the most critical work in the company. You do not make the most important decisions, and you do not determine the success of the company. Others do all of that work, and *they* should be recognized as the heroes when it all comes together. As the CEO, however, you do set the direction, frame the plan, determine the culture, and, as explained in this step of the ALIGN Process, cheer the team along. This is your journey. Every positive step in the right direction deserves recognition. Every milestone reached means success for the entire organization and so should be treated as a success. If it's worth doing, then it's worth doing well, and it's worth recognizing work well done. In my experience, teams that celebrate their wins together move forward the fastest.

I once worked on a project with a manufacturing company in Asia. I needed one of the frontline employees to take the lead on a project, and we worked together for about six weeks. Each week, I did a review with the leadership team and shared with management the progress that had been made. The discussion would always start with my explanation of how the lead frontline employee had done an incredible job of listening and applying the strategy in his team.

After only the third week, I got this inevitable comment and question: "I never knew we had a frontline employee with that kind of talent. Do we have any more like that?" By then I had heard so many versions of that same comment over the years. I simply said, "Yes. Many." Although I had given the employee we were discussing some training, what I had mostly done was challenged him. I had asked questions, stood alongside him, and let him know that he could have a huge impact on the company. And then, as he showed that he was willing, I recognized his achievements – to him and to the leadership team. That employee was subsequently promoted and then asked to lead a new production line as it was installed.

There are a lot of good reasons to be happy in business. Find them, celebrate them, share them, and enjoy them.

Work with Heroes

If you want to work in a company with heroes, make the people around you heroes. It's not hard to do. First, find out what someone is good at, then help them get better, and then you recognize their growth. Then... you have a hero.

In all the work I do to improve manufacturing companies, it's never me who's making a product, making a sale, doing the quality testing, doing the production planning, or designing new products. The people who do those things are the ones who get the results.

When a company improves in the desired metrics by 40%, which is common in my experience with the companies and clients I've worked with, then that 40% is due to the frontline employees. To achieve a 40% improvement, there was a lot that had to go right, and so that's a lot of people to give credit to. If you're looking for them, you will see that there are miracles happening all around you, every day. Someone goes out of their way to help a colleague; someone looks for a new solution to an issue; someone uses their skills to clean and maintain the equipment; someone helps a customer understand how the product can transform their life. "But aren't those just regular day-to-day activity?" you may ask. Absolutely. And within in the midst of those daily activities are heroes who go unrecognized every day.

When you recognize positive behavior, it breeds more of the same. Recognition is powerful and it's free to give. Help the people around you become heroes, and you will soon be working in an heroic, successful organization.

Catch People Doing Things Right

When kids are young, we tend to notice the things they do well. The first time they can point; the first time they grab for something; the first time they lift food to their mouth. As they try for success, we cheer them on. We forgive the mistakes they make, celebrate each attempt to improve, and cheer when they finally make it. We don't criticize and ridicule every time they stumble and fall. We stay positive and focus on successes.

And then, at some point, we switch our focus to criticism. "Why couldn't you do it better?" "Why do you keep making mistakes?" "What's wrong with you?" Somehow, our expectations increase beyond the child's current abilities, and we begin to notice all the negatives. This is part of the reason why many kids struggle in school and why improvement slows down after those rapid-growth first years.

By the time people have been working for a few years in a business, they are often starved for recognition, praise, and support. Their attempts to improve and their success can be switched on again rapidly as leaders pay attention and focus on the positives.

Reward Practice not Perfection

In order to improve, we need to make mistakes. Time needs to be given for practice, but this is often overlooked in a manufacturing business.

In motor racing, the pit crew can change tires, fill the fuel tank, change visors, and do a dozen other activities in a matter of seconds. This expertise does not come only from trying harder during every race. It comes from continual practice, striving for personal best results, reviewing every movement, and making decisions for improvement.

The same is required in a manufacturing department. If you want a production team to become masters of line changeover, they need to practice line changeovers. If you want a sales team to be masters of negotiation, they need to practice negotiation. We do analyze and review results after certain milestones, but we don't need to wait for those. Schedule time to encourage the teams to become masters and to encourage practice. There is a cost to doing this, but a return on the investment is inevitable. Training and coaching the team through to perfect practice is even better, but *any* practice will put you ahead of 90% of your competition.

Recognition is Better than Rewards

I hear you when you say that you can't afford to keep giving out awards for this and that. You may be wondering if giving out too many awards demeans the whole thing, or if adding the reward cost to the workers' salaries would be enough to make them happy, since that would be simpler.

In the section above, I purposely don't mention *rewards*. I am talking more specifically in this step about *recognition*, and I have never seen anyone be "too recognized." I am also a fan of rewards – small, thoughtful gifts specifically chosen to honor an outstanding achievement

– but I will address that issue later. In my experience, recognition trumps rewards.

Telling someone, "You did a good job" is nice, but it's not recognition in the way I'm talking about. Effective recognition – meaning recognition that benefits all involved, including the company – includes several elements:

- Identifying the achievement you're recognizing the individual for
- Identifying the specific behavior that allowed them to make this achievement
- Knowing how this achievement helped to take the business forward in the right direction
- An expression of genuine gratitude
- A reinforcing of company direction

For example, when you see the cleaning guy cleaning the restrooms and doing a great job at it, you can say any number of things relevant to the elements above to recognize his contribution:

- "I just want to say that these restrooms are spotlessly clean and are some of the best-maintained factory restrooms I've ever visited."
- "I saw the way you were cleaning underneath the sinks with a hand cloth. I rarely see anyone paying such attention to detail while cleaning a restroom, and it's nice to see."
- "Having a high-quality, well-maintained restroom really shows the employees that we care in this company, and the way you treat the people who use these facilities is the way we should all be treating our customers."

- "Thank you for your contribution to making this company the fastest-growing widget company in southeast Asia."

That took 30 seconds of your time. 30 seconds! Do you really think people hear thoughtful, appreciative comments like that too often?

If you still wonder if this will really work, or think it's just a cheesy manipulation of the employees, try it at home. For 30 seconds while your spouse is preparing dinner, while you're taking the kids to football practice, or taking the dog out for a walk, recognize your family members for their contributions

When was the last time someone said something to you to recognize and appreciate you? How did it feel?

As the setter of company culture, your every move is followed, scrutinized, and –eventually – copied. Be brave enough to start some habits of your own that will spread like positive wildfire through the company.

Award with Experiences

I do also like rewarding people, but I prefer small awards. And I prefer awards that are experiences rather than objects. Experiences stay with us, and great experiences can last a lifetime. I also like awards that involve the maximum number of people involved, so that all can feel great about their achievements.

In one company I worked in, we used to celebrate every time a production line achieved a new personal best result in one of the priority areas. We used the term *good runs* for these instances, and the metric we tracked was how many good runs were achieved in a certain time period, such as during a shift, a day, a week, or a month. Each time the team managed to achieve a new best result, there would be an ice lolly for everyone in the cafeteria for lunch. We would also put up a sign to tell which team had achieved the new benchmark, what their learning was that allowed them to achieve it, and how pleased they were to be the ones donating the ice-lollies. It was a low-cost, high-impact, experiential awards system that worked!

In another company, we allowed the winner of an award to "donate" an experience to a person in need. One of the most touching and heartwarming rewards for the entire company was to be involved in such a moment of emotional happiness when the donation was given. One award winner chose to experience taking toys to a local blind school, and doubled the impact of the donation by allowing the wife of a colleague, who had recently suffered a difficult trauma, to make the trip to give the donation. We made a video of the experience so that the entire organization could watch and be involved. This provided a big impact from a small award, and shared positivity and joy.

There is a place for well-considered awards. Make them special for all.

Action Items for Step N

Nurturing feedback and recognition is possibly the easiest and most impactful change to make in an organization. I rarely see a company that is using recognition well. There are rarely award nights, celebrations, or recognition events in manufacturing companies. The fact that that is out of the ordinary and unexpected is even more reason for you to distinguish your company by leading the way.

Answer the following questions to give yourself a start on this process:

- How will we give feedback to the entire organization?
- How will we encourage leaders in the company to embrace this method?
- How will we celebrate successes?

Bonus Materials

This section is intended to give you further examples and some guidelines to help with your implementation of Step N: Nurture Feedback and Recognition.

Rah-Rah vs. Ho-Hum

It is simply not in my character to be the rah-rah person – the person who organizes company events, gets everyone up on stage, buys the birthday cake for the team member, or arranges karaoke sing-alongs on the bus. It took me a long time to realize how important that person is to the organization, but because it's not part of my character I simply couldn't see the value of it for a long time. I used to wonder, "What do people get out of this?"

To me, there were a surprising number of people for whom family-oriented socializing activities were very important, critical even, in order to feel happy and fulfilled at work.

I once had a personal assistant who was good at her job but needed to leave after a couple of years. I asked her if she had been happy working for me and she said, "Yes, except that you've never acknowledged my birthday." I was stunned. I had fallen on that point before, and that had prompted me to institute a system whereby I recorded everyone's birthday and tasked a person on the staff to buy the cake and lead the birthday recognition. Who was that person, and why had they missed my PA's birthday? Well, of course it was my PA. The one birthday celebration she did not want to arrange was her own!

If you are a rah-rah person, that is a great asset. If you are not, then find a team member who can play that role more naturally.

Outcomes vs. Certification

I hear company leaders say, "We need to get ISO certification," but when I ask them why, their answer is usually something to do with marketing. They feel that the certification is a requirement rather than part of a support system. This goes not only for ISO certification but for any type of certification. We live in a society that often puts more value in the piece of paper than the actual skill set. Let's look at this issue in more depth.

If a student goes to university with the specific goal of *getting a degree in physics*, there will be a certain level of engagement between the student and the material. *Getting a degree in physics* is a goal that can be carried out in a variety of ways.

If, on the other hand, a student goes to college with the specific goal of learning all she can about physics in order to become an expert astrophysicist, the level of engagement will be significantly different. In this case, the piece of paper, the certificate showing that she has earned a degree, will more likely be a byproduct of the journey, because it is not as important as the focused expertise she gained and the increased options she has gained.

I see the same type of thing happening in companies around certification. If the company desires to release exceptional quality products to the consumer, they will learn all they can about creating a manufacturing and supply chain environment that assures quality. With that goal in mind, getting required certifications will be easy.

The same approach goes for safety, employee welfare, and any other area that is now regulated with an expectation to be certified. Focus on the outcome you want to achieve, put the systems in place to ensure it, and the certification will be a certain byproduct.

Changing Behaviors vs. Changing People

We've talked a lot about catching people doing things right and recognizing their contributions. There are other times, however, when you will catch people doing things wrong. I'm not talking about mistakes, which are to be expected and, in some ways, encouraged as part of the learning process. I'm talking about people knowing that something is the wrong thing to do but doing it anyway. I like to think that people are mostly good and are just trying to build a better life for themselves and those around them. So if someone does something wrong, then it is their behavior in that particular situation that we don't like, not the person. It is, therefore, their behavior that we want to change.

Just as we had a method for recognizing strong performance, here is a formula I like for adjusting a person's behavior:

- Make sure that there is no doubt that the behavior happened (you can even ask the person, if you need to).
- Find time alone with the individual (while recognition is often effectively expressed in public, reprimands should be dealt with privately, whenever possible).
- Let the person know the specific *behavior* that was not acceptable and why.
- Ask the individual why the behavior happened, in this case, but do not excuse it.
- Inform them that the behavior must not be repeated or there will be additional consequences. Be clear about what those consequences are. Let them know that you do not want those consequences to happen, and so you will support them however possible to ensure that they don't.
- Let them know you appreciate their contribution, and you appreciate them as a person.

There should be absolute clarity for both of you regarding what the reprimand is for. And there should be no room for an excuse. The whole discussion should not take more than a couple of minutes.

Let's look at an example. You're walking along a production line and a frontline employee is reaching inside one of the machine guards when the equipment is moving, which is against the work rules. Here's an example of what you could say as you apply the formula above:

- "Let's step back away from the equipment for a moment so we can talk."
- "I just saw you reach inside the guard with the equipment moving, which is against the work rules. Is that what happened?"
- "Putting your hands inside moving equipment is this industry's number one cause for workplace accidents. That work rule is in place to ensure that our employees never get hurt. It's not acceptable to me for this work rule to ever be broken. I want to keep you safe and well, and that's why we're talking right now."
- "Why did you feel the need to break the work rule on this occasion?" Potential answer: "The machine was dirty, and I didn't want to stop the line to clean it, as that would take additional time, and we would lose production."
- "Safety is a priority over all other aspects. The work rules are there to protect you and keep you safe. We will look together at how to solve the contamination issue, but I never want you to do this behavior of reaching into a moving machine again. This is my verbal warning. It will be noted in your records, and if you break a work rule again this year you will get a formal written warning. I never want you to be given a warning, but – most importantly – I never want you to take an action that could give you an injury. Is that clear?"
- "You are important to this team, and you have great skills in operating this equipment. We need you here and we need you safe."

Even with the responses of the individual, this should not take more than a couple of minutes.

If at any stage of the encounter, you get a highly charged emotional response, ask the individual to step away from the current circumstances to talk further. Move to a quiet and calm environment, and get ready to listen. There may be a sensitive undercurrent that requires deeper exploration. In that case, asking a neutral colleague to observe the discussion may help.

You are responsible for the standards you set. Whatever you are willing to walk past sets a standard. Keep the standards high, and be confident. If the right standards are set in order to take the business forward, then you should be proud of enforcing them and giving everyone the maximum chance of business success.

Relationships vs. Pay

In study after study, there is one main reason that people leave a company. It is not the pay, the working environment, or their work-life balance. It is their relationship with their immediate boss.

People will work for the right boss in the most unusual circumstances and conditions. You can't pay people enough to stay with the wrong boss.

If you are reading this book, it's likely that you are a leader, even if you're not officially titled as one. There are people around you or working for you who watch your every move. Your mood, your behavior, and your actions all have an effect on them and make their working day a joy or a misery.

As you communicate with those around you, as you set the standards and maintain them, pay attention to the impact you're having on others. If your actions are moving the company and the employees in the direction the company requires, then you are playing your role as a leader. If you can do this with a high level of love and respect for those around you, the chances are you will also be an inspirational leader.

Fit vs. Frazzled

My number one priority in life is mental and physical fitness. It is a higher priority than family, finances, work, or friends. There is a simple reason for this. When I am fit in mind and body I am in a condition to help and support others. I can give them the very best of myself by being calm, present, and healthy.

When I am mentally or physically unfit – stressed, distracted, tired, sick – I am a burden, not a help. At those times, it is better not to be around or in contact with people, let-alone try to lead them or make decisions.

This is a beneficial stance to support others to take, as well. Employees are better for the organization when they are mentally and physically in their best condition.

Fitness requires ongoing attention and effort, and I consider a key support role of the organization to be helping employees stay in optimum mental and physical condition.

When we attend to the fitness of our manufacturing equipment, we:

- Listen to hear abnormalities and take corrective action
- Keep it well maintained
- Set it up to the optimum conditions
- Give it the right environment to ensure a long life and quality output
- Spend money and time on upgrades
- Provide the right fuel

Just as for our manufacturing equipment, the leader's job is to set the conditions for the employees to be able to give their very best contributions and to continually increase their capabilities. Consider what is required to achieve this, and set it in motion.

Understand vs. Defend

There are seven billion potential critics on the planet. The more of them we come into contact with the higher the possibility of being criticized. It is almost inevitable that both you and your organization will, at some point, be criticized. At times this may seem fair and at other times the criticism will seem completely unjustified.

I like to view criticism like I view any other problem. The first response should be to seek more data, to gather more information about the problem.

When you ask for more data, more detail, more explanation, and seek to understand, the critic will be helping to solve an issue. At no point is this about winning, so there is no need to defend. There is no winner in a war. In the process of fighting a war, everyone loses. So seek to understand, because in every criticism is a nugget of learning. Learn, but don't be distracted from the company's direction.

When BP was criticized for leaking oil throughout the Gulf of Mexico during the Deepwater Horizon disaster, there was very little effort on behalf of then-CEO Tony Hayward to listen, seek to understand, collect more data, and *then* respond. After his defensive reaction to the incident, that CEO didn't last long.

Summary of Step N

Step N is all about feedback. Without feedback there is no concept of progress. Without recognition there is little incentive for improvement. Put a structure in place to catch people doing things right, to appreciate progress, and to celebrate the milestones along the way. Enjoy the journey and encourage others to enjoy it too. In business, there is no final destination, therefore the journey is everything.

---○---

CHAPTER 8

---○---

Possible Roadblocks on the Journey

"If you're trying to achieve, there will be roadblocks. I've had them; everybody has had them. But obstacles don't have to stop you. If you run into a wall, don't turn around and give up. Figure out how to climb it, go through it, or work around it."
Michael Jordan

No journey is perfectly smooth. Due to the high complexity and variability in manufacturing, there will always be situations that we are not fully in control of and that need some maneuvering to get over, under, around, or through. If you have set up your organization according to the ALIGN Process Steps, you will be in good shape to overcome obstacles, by using your reviews, your listening skills, and supporting people and teams in their very specific roles and responsibilities. These obstacles and what is learned from them can eventually be turned into materials for on-the-job trainings and the development of additional skills, which can then be translated into a method to stop the incident from ever happening again.

I have listed below some of the common obstacles, in case you need to navigate around them. My comments below may seem oversimplified, but I believe these obstacles, and how they are dealt with, offer pivotal turning points that can help you decide who you want to be in your role as the person who leads your organization toward success.

Remember that you are not alone. You will also, at times, need to ask for help and support.

"I don't know how to create a direction I am passionate about."

Without passion you cannot be an inspirational leader. Dig deep. There is a reason that you are where you are. Something about how and why you got there compels you and encourages you to keep moving forward. Your passion is already within, so focus on allowing it to emerge into view.

"My passion, desire, and direction for the company don't align with what the chairman or board of directors expect from me."

If they are not swayed by your discussions on this topic, and you are not aligned to their direction, then it might be time to move to a different company. There is nothing more soul destroying and confidence sapping than spending a large portion of your life on an endeavor you are not committed to.

"Some of the long-serving leaders will not get on board with this process. However much we discuss it, there are signs of a rebellion."

A successful organization requires all the people involved to be rowing in the same direction. Those who insist on rowing in a different direction need to get out of the boat – and you need to ensure that they do.

"How do I stick with this so the process isn't viewed as the latest fad?"

Schedule it. Your schedule reflects what you believe, your real priorities. When you schedule and commit to the 5 Steps, you will see the progress, and that will eventually be seen and believed. It will be messy at times, but your consistency is the key. This is your process, so make it work.

"Some groups are resisting following the process – especially when it comes to reducing their budgets based on the priorities."

The purpose of Step L is to gain alignment and commitment. This is the stage for inviting debate and then making decisions. Following the debate, commitments are made. *Sticking to commitments* is one of the ground rules from Step A. Ground rules are non-negotiable. If ground rules are not being followed, you need to take action to show your commitment to them.

"We're afraid to celebrate today in case the results are worse tomorrow."

Organizations based on fear can operate this way. Imagine if the soccer team that won the World Cup didn't celebrate because they weren't sure if they would be able to retain the trophy at the next tournament four years later! Celebrate each success that's moving the company in the right direction.

"Someone in the leadership team said there's another process that is better for us to follow."

There are as many opinions as there are people. Your role is to discuss the options, make a decision, then follow whichever process you chose. The worst decision you can make is no decision. Make an informed decision, and then make it work.

"How can I do daily video messages when my global workforce speaks many languages?"

Use translators.

"How do I make the leap of faith required to allow my team to take over making more decisions?"

Set the direction, leap, review, adjust, leap again.

There are going to be obstacles. Don't get fazed by them to the point of moving off course. Keep taking one step at a time on the path toward your destination. Lean back, breathe, and relax. You've got this!

---○---

What Do We Do Now?

"To be always intending to make a new and better life but never to find time to set about it is as to put off eating and drinking and sleeping from one day to the next until you're dead."
Og Mandino

There is significant content in this book, but much of it is food for thought. I have included many examples to help you think about the subject matter, but the specific process for transformation is relatively simple: Follow the 5 Steps of the ALIGN Process, write down your decisions, and start on that path. Overthinking will not help this situation. Your team is waiting for your leadership. They are waiting for you to blaze a trail for them to follow. They are looking for your direction, your passion, and your inspiration to drive them forward.

They will need your help, support, and encouragement on the journey. It's all about the journey.

As the CEO, you are forbidden from making this difficult. People cannot follow difficult instructions or go in confusing directions. The process must be kept simple. Tell your people where you are all going, how you will get there, your expectations for the journey, and how you will support them. Allow them to make the decisions that they are better positioned and more capable of making than you are. Find the positives as you take action. Recognize and celebrate each step forward in the right direction, and cheer your team on.

This ALIGN Process works in real life. It is tried and true. It is compiled from the best practices of the leaders who have many of the most successful manufacturing businesses in the world. I have decades of experience in applying it and seeing it work wonders in a great variety and number of manufacturing facilities. It will work for you, and I am here to give you support.

If you choose to go it alone from here, simply let me know how you progress. I'm interested in how you do. Stick to the path and kindly allow me to share in your celebrations.

Should you need a sounding board along the way, feel free to look me up. I always have an ear for those who are positively transforming the world.

REFERENCES

Kouzes, James M. and Barry Z. Posner. *The Leadership Challenge*, Jossey-Bass, 4th ed., 2007, p. 30.

Csikszentmihalyi, Mihaly. *Flow: The Psychology of Optimal Experience.* Harper & Row, 1st ed., 1990.

ACKNOWLEDGMENTS

Getting the nuggets from 30 years of manufacturing experience down on paper is no simple feat. There were many steps from when the concept first popped into my head to the finished book and, as I learned during the writing process, most potential authors drop out somewhere along the road before their book is done. I couldn't drop out! This message is far too important. I feel for the millions of people involved in manufacturing, and they deserve a chance to find their roles rewarding, enjoyable, and meaningful.

I want to thank Robin Sharma, who encouraged me to share my message. He has done so much to inspire leaders with his many books including *The Monk Who Sold His Ferrari* and *The Leader Who Had No Title*. When I told Robin I wanted to write a book specifically for manufacturing businesses, he encouraged (challenged) me to do so. Challenge completed! I'm grateful to him for being a disciplined mentor and role model.

Without my wonderful experiences around the world that P&G allowed, I would not have had the foundation I did to help make a difference. There are so many people to thank within P&G, as well as associated suppliers and contract manufacturers, who taught me so much during our experiences together. I thank them for listening and for showing me what really does and doesn't work.

I also want to thank Angela Lauria of The Author Incubator, for making the book-writing process friendly in the eyes of an engineer. Angela's wonderful guidance at every stage allowed me to share my message in a way that has a better chance of making a difference.

To the Morgan James Publishing team: Special thanks to David Hancock, CEO & Founder for believing in me and my message. To my Author Relations Managers, Megan Malone and Niara Baskfield, thanks for making the process seamless and easy. Many more thanks to everyone else, but especially Jim Howard, Bethany Marshall, and Nickcole Watkins.

There is no doubt in my mind that a journey of lifelong learning was instilled at an early age. For that gift, and so many more, I want to thank Irene and Allen Snook who were role model parents and who continue to inspire me every day.

Finally. I want to thank my wife and best friend, B. She has brought such joy into my life, in so many forms, and allowed me to continually expand my boundaries. I am truly grateful.

ABOUT THE AUTHOR

Kevin Snook is an advisor to manufacturing CEOs around the world, helping them transform their businesses and deliver breakthrough results by giving frontline employees the information, tools, and capability to make the very best decisions every minute of the day.

As a manufacturing leader in P&G for 17 years, Kevin had hands-on experience of producing billion dollar brands such as Pampers, Always, Gillette, Pantene, Cover Girl, and Head & Shoulders.

After successfully growing P&G's Contract Manufacturing Division for Asia, Kevin left P&G to become managing director of a 4,000-employee manufacturing business. The business grew by implementing many of the best systems Kevin had formulated.

In 2010, Kevin felt the calling to help more manufacturers and so started his own consulting business, Saxagon Limited (www.saxagon. com), to use the best practices he had discovered and developed to transform manufacturing companies around the world.

Through his 30 years of service, from frontline employee to CEO, Kevin has experienced and worked through a vast array of challenges for manufacturing companies. In the process of striving for and learning from the root causes of each equipment breakdown, quality incident, safety accident, initiative issue, and delivery failure, Kevin always sought permanent solutions and systemic change that eliminated the chances of such an issue happening again. Reviewing, analyzing, and learning from how things work during the best of times have given Kevin unique insight into what works.

As an entrepreneur, founder, and consultant to hundreds of manufacturing companies across more than 25 countries, Kevin has a deep sense of what it takes to deliver the magic combination of growth and stability in manufacturing companies. And he has gone further, by learning how to make that outcome simple.

Born in England, Kevin relocated to Asia in 1996 for a two-year assignment. More than 20 years later, having experienced first-hand the rise of "Factory Asia," Kevin still resides in Asia with his family, and considers it a comfortable base for his global businesses.

Website: www.makeitrightbook.com

Email: kevin@makeitrightbook.com

Facebook: www.facebook.com/KevSnook

THANK YOU

It can be difficult for CEOs and business leaders to know where to go for support. I know – I've been there!

This book's step by step guidance is only the beginning of the resources available for your ongoing support.

As a thank you for picking up this book, I invite you to access the following resources, for readers only:

- A printable checklist of the five ALIGN Process steps, with a summary of each step.
- Four sample two-minute CEO daily videos, and a guided script to follow for effectively communicating with your team.
- A one-page check-sheet for handling any review: How to Run Effective Meetings

To collect these resources, visit www.makeitrightbook.com/bonus.

Wishing you ongoing success and a highly rewarding manufacturing life,

Kevin Snook

Morgan James
Speakers Group

↗ www.TheMorganJamesSpeakersGroup.com

We connect Morgan James published authors with live and online events and audiences who will benefit from their expertise.

 Morgan James makes all of our titles available
through the Library for All Charity Organization.

www.LibraryForAll.org

9 781683 506706